AGAINST THE GRAIN

AGAINST THE GRAIN

How Farmers Around the Globe Are Transforming Agriculture to Nourish the World and Heal the Planet

ROGER THUROW

A MIDWAY BOOK
AGATE
CHICAGO

Left: *Degraded to Flourishing*
Terraced farmland and diverse cropping system in Ethiopia

Printed in the United States
First printed in March of 2024

10 9 8 7 6 5 4 3 2 1 24 25 26 27 28

ISBN-13: 978-1-57284-340-0 (paperback)
ISBN-10: 1-57284-340-3 (paperback)
eISBN-13: 978-1-57284-885-6 (ebook)
eISBN-10: 1-57284-885-5 (ebook)

Cataloging-in-Publication Data is available from the Library of Congress

Cover photo and all interior photos by Anne Thurow

Midway Books is an imprint of Agate Publishing.

Agate books are available in bulk at discount prices. For more information, go to agatepublishing.com.

For my brother Tom, who loved all creatures great and small and the soils that nurture us all.

CONTENTS

PROLOGUE

Why Would I Do That?

On the horizon, I am told, is a lush paradise, a garden of spectacular abundance. It is hard to believe, for all around is a moonscape of largely denuded plains and hills. We are skimming across Ethiopia's Great Rift Valley highlands on a narrow dirt road, a ribbon of dust winding through fields that have far too often yielded only hunger. Eerie sand-and-dirt sculptures rise beside deep gullies, the scars of miseries past crafted by years of wind and rain, nature's relentless artisans of erosion.

In the distance, suddenly, a shimmy of life. Is it a mirage, or the promised oasis? A cluster of trees comes into view. And then a small house. There's a fence fashioned from sticks and dried cornstalks. A tall, thin figure saunters to the gate. A miracle worker on this bleak landscape.

A farmer.

Left: *Ethiopian Gothic*
Abebe and Tsehainesh harvest fruit on their Great Rift Valley land

"Welcome, welcome," says Abebe Moliso, enthusiastically bumping shoulders with his visitors in the traditional manner of greeting. "I have something to show you," he tells old friends from the United Nations' World Food Program. "This land was dead. Nothing would grow. And now look! Come. You will see."

Abebe, a most hospitable man, spreads his arms wide like the branches of the trees in front of his house, savoring the glorious shade. He selected these trees for their broad leaves, and the canopy of relief they provide from the searing sun.

He leads the way to the side of the house. There, around the corner, begins a row of fruit trees he sowed as a green hem to mark the edges of his land, a four-acre rectangle. Mangos, avocados, papayas, bananas, and apples hang from bended boughs.

"Apples?" one of his visitors asks.

"Yes, I too was surprised. I didn't know apples grow in Ethiopia," Abebe admits. "Then I saw them on another farm. And I asked to have a few seeds to plant myself. Now I have apples."

Behind the house is a kitchen garden, an enticing patch of nutritious produce: carrots, cabbage, beets, and tomatoes, and a variety of herbs. Abebe waters them from a small pond that suddenly emerged a few years back from an underground spring he didn't know existed until it gurgled to the surface. He introduced water lilies to the pond to provide cover from the sun and limit evaporation, and he encouraged algae to grow to attract dragonflies and bees. And when the bees arrived, he built a hive to harvest honey.

A series of narrow plateaus descends like a staircase from the garden. Abebe shaped the naturally sloping land into these terraces to keep the precious rainwater from rushing away, and to hold the valuable topsoil in place, to slow the erosion. On the plateaus, he planted crops. Beans, peas, Irish potatoes, sweet potatoes, barley, teff, millet, cassava, coffee, and peppers—green and red and chili-hot.

Between the plateaus he nurtured an organic ecosystem, introducing bushes that produce natural pesticides and trees that shed their leaves as the crops grow, assisting the plants by retaining moisture on the ground and adding crucial nutrients and carbon to the soil as the leaves decay.

The terracing gives way to a meadow hosting rows of maize. Beyond that is a stretch of tall grass where a dozen dairy cows graze. A stream meanders past, forming the bottom border of his farm.

It is an overall tidy place, but Abebe encourages a bit of mess. The leaves aren't raked and the stalks and stems of the crops aren't collected after harvest; they remain scattered on the ground to shield the soil from the sun and enrich all the unseen microbes in the underworld that nourish life above ground. He also tolerates weeds—nurtures them, actually—to help attract the insects that work the soil and the pollinators that bring an array of plants, vegetables, and fruit to life.

Everything here has its purpose and place.

This farm and its riotous diversity is remarkable, I tell Abebe. But surely, cultivating such diversity must take a lot of time and effort. Had he ever thought about putting all his acres into one crop, like maize, a staple food for much of Ethiopia and Africa?

He fixes me with a quizzical look. "Now why would I do that?" he asks.

Such monocropping has become a hallmark of modern farming around the world, of the determined push to boost commodity production as a way to maximize farmer effort, yields, and income.

"No, no, no," Abebe says, waving off the idea. "We did that

once, too, growing mostly one crop, year after year. And then nothing more would grow. It ruined everything."

His family and neighboring farmers, he explains, had grown mainly maize one season after another on the same plots of land, following the global agriculture orthodoxy. Over time, such repetition sapped the soil of the same nutrients, without adding new ones; chemical fertilizers, spreading in use across the world, were too expensive for him and many of his neighbors. As it depleted the soils, the monocropping deprived their families, too, of a nutrient-diverse diet. It made a steady income more precarious, tying the farmers to the fluctuating price of a single commodity. If pests or disease came for that one crop, they could lose everything. Still, they persisted with the practice. The maize yields from the depleted soils became smaller and smaller. The more the farmers worked to grow something, the less they actually grew. Trying to eke out a bit of food from tired soils that had nothing left to give seemed futile. Similarly, their small herds of cattle grazed the same patches of land, year after year, to the point that all the grasses were gone. In an effort to increase food production, the farmers cut down nearby forests, slashing and burning, to clear more land for planting crops and grazing cattle. But this was just as counterproductive. As the trees and forests—drivers of the precipitation cycle—disappeared, rain became ever scarcer. When it did come, very little of the water was absorbed by the barren, sunbaked land; the majority rushed away. Over time, the entire ecosystem dramatically changed. Water tables sank, streams dried up. Bushes and grasses shriveled and died. Temperatures rose, hot winds withered. Birds flew away, as did the butterflies and bees and other pollinators. Droughts increased in frequency. The hunger season, the time of profound food deprivation between harvests, lengthened. The farmers stood in line to receive food aid for their families, their survival dependent on other farmers elsewhere in the world. Hungry farmers—the absurdity of it. Their children were malnourished, stunted and wasting away; many of them died. Famine loomed.

Abebe seems exhausted by reciting this history. "We did it to ourselves," he says.

Eventually, the Ethiopian government and humanitarian agencies working in partnership with the local farming communities fashioned a strategy that restricted the use of the land. Abebe was just a young man then, barely twenty. The erstwhile farmers agreed not to cultivate crops or graze any cattle, and to stay out of the forests. No more cutting down trees. No more squeezing the soils. The land must heal. The ecosystem must be restored.

The farmers would be paid in food, provided by the World Food Program (WFP), for their labor to help with the rehabilitation. The WFP, charged with feeding the world's hungry, feared that standard food aid handouts to farmers would need to go on forever if there wasn't radical change to agriculture methods. That's when the land terracing began. The farmers first planted grasses, small shrubs, and tree saplings on the plateaus instead of annual crops, to renew the vegetation and soil. Instead of plows, they were given shovels to dig shallow pans to collect any rainwater so it could soak into the soil rather than run off, thereby restoring the underground aquifers. After several years, the springs resurfaced and natural ponds formed. As the vegetation reappeared, so did the pollinators and the wildlife. Life returned to the land.

And so, too, did farming.

As he resumed cultivation, Abebe conceived a new philosophy: experiment, adapt, question, challenge. "Why would I do that?" became his mantra.

Why indeed would he go back to the old practices, even if

that was how other farmers around the world worked? Abebe, now in his midforties, says, "We saw what happened when we relied on one crop, when we didn't change. We have learned that it isn't wise to plant only one type of crop. It's too risky."

He questioned and learned from a new class of agriculture advisors dispatched by the Ethiopian government and organizations like the WFP. He determined that he would grow many crops instead of just one, to regenerate the soils, to diversify diets, to spread the risk. He learned to work with nature rather than against it. If he was good to his land, he believed, his land would be good to him. And his modern-day Garden of Eden grew where once nothing did.

"Now I have crops coming ripe all year long. If one fails, another succeeds. We have a steady flow of food and income," Abebe says. "We have realized the fruits of our work."

Farmers from near and far began visiting to see what he had done. His farm has become a model for his neighbors and his entire country. He wants everyone to succeed on their farms, to finally banish malnutrition and hunger from this land.

As he returns to the front of his house after the tour of his crops, Abebe eagerly greets a group of local women gathering under the shade trees. They have formed a savings group to pool income from their kitchen gardens. First, they saved for school fees for their children; then, they saved to buy sheep and goats. Abebe's wife and farming partner, Tsehainesh, hosts nutrition and cooking classes to talk about the benefits of vegetables and fruits and diversified diets. Together, the women marvel at their families' health improvements; instead of weeping over children suffering from malnutrition, they now celebrate graduations from high school and college. There has been a collective graduation, from barren land to improved nutrition to better education.

Now we follow Abebe into his home. On a living room wall hangs a diploma from his son, the first to graduate from college. He again opens his arms wide, envisioning a wall full of diplomas. Here, learning and questioning has pride of place.

"We have seen the mistakes of those who have come before us. Our dead land is living again," Abebe says. "God gave me an open mind to learn. And I hope I can open the minds of others."

◈

More than a half century ago, before Abebe was born, American agricultural scientist and crop breeder Norman Borlaug was awarded the Nobel Peace Prize for sparking what came to be known as the Green Revolution with a new variety of wheat plants that boosted global production and reduced famine in many regions of the world. The Nobel committee, in 1970, praised Borlaug for defusing a grave threat to humanity by accelerating the pace of food production ahead of population growth, saving countless lives from starvation. "In this intolerable situation, with the menace of doomsday hanging over us, Dr. Borlaug comes onto the stage and cuts the Gordian knot," the committee said. He had "turned pessimism into optimism in the dramatic race between population explosion and our production of food."

The Gordian knot: an intricate, seemingly intractable problem, a challenge of mythological proportions. In the 1960s and '70s, the Green Revolution's zealous quest to increase agriculture production helped overcome famine in India and Pakistan and other parts of Asia where starvation had been immense and unrelenting. Agricultural progress, propelled by advances in science, technology, research, and investments, helped to shrink global levels of extreme poverty by raising farmer incomes and reducing hunger.

All the while, though, a new Gordian knot has been forming, born from the Green Revolution and the modern agriculture system, with its reliance on chemical fertilizer, irrigation, and farmland expansion, that rose in its wake. The very agricultural actions intended to nourish us also endanger our environment and biodiversity, and, paradoxically, our health. The original dramatic race—between an ever-growing, ever more prosperous population and the rate of food production—persists. But now a second dramatic race appears, between humanity's two most pressing imperatives: nourishing the planet and preserving the planet.

The fear now is not so much that we will run out of food, but that we are dooming ourselves and our planet by the way we grow this food. The essence of today's challenge lies not in *how much* we grow (the mission of the Green Revolution), but in *what* we grow and *how* we grow it to satisfy shifting diets and meet the daily nutritional needs crucial for healthy individual and societal development.

Cutting this new Gordian knot requires substantially increasing food production while at the same time drastically shrinking agriculture's impact on the environment. World population reached 8 billion in November 2022 (up from about 6 billion in 2000) and is expected to reach 9.7 billion by 2050 and possibly peak at 10.4 billion in 2100, according to the United Nations' rapidly spinning population clock. At the same time, the global middle class, with its higher consumption patterns, has significantly expanded to include about half of the world's total population. It is forecast by some measures to continue to grow to more than three-quarters of the population by 2050. This large upward economic migration is good news. But it puts increased strain on the global food chain, and greater demands on agriculture. The World Resources Institute (WRI) estimated the demand for grains, cereals, and other plant crops is projected to grow by more than 50 percent by 2050 (from 2010), and the demand for meat and dairy to grow by nearly 70 percent.

To keep pace with these projected increases in food demand by midcentury without clearing additional land for crop production that would further damage our environment and biodiversity, the world must boost productivity of both crop and livestock systems by those same amounts on existing agricultural acreage. If not—if agricultural yields remain frozen at 2010 levels—agricultural land will likely need to expand by more than seven billion acres by 2050 (from 2010), the WRI calculated. And, even if yields continue to grow at roughly the same annual rates of the past few decades, cropland and pasture area would still likely need to expand by 1.5 billion acres by midcentury.

Accomplishing this—producing more food to nourish the population while at the same time protecting the planet from the consequences—will require a total re-engineering of our post–Green Revolution food system.

Satisfying the relentless human appetite for food, the WRI and other researchers noted, has been the biggest driver of land-use change throughout history. Rather than working in harmony with nature, agriculture has often sought to bend nature to its will. To feed the planet's population, worldwide agriculture has already cleared or converted at least 70 percent of grassland, 50 percent of savanna, 45 percent of the temperate deciduous forest, 27 percent of tropical forests, and 50 percent of the wetlands. Agriculture uses 70 percent of all fresh water globally. Since the 1960s, almost 1.5 billion acres of forests and woody savannas globally have fallen under the plow. At the same time, about one billion acres of farmland have been abandoned due to degradation, largely from over-farming or over-grazing, according to a study from Stanford University. The UN's Food and Agriculture Organization estimates that about 35 percent of the earth's soils are already degraded. And Professor Rattan Lal of Ohio State University, winner of the 2020 World Food Prize for his pioneering work on soil conservation, calculates that the world's cultivated soils have lost up to 70 percent of their original carbon stock, which contributes to soil fertility. Soils form one of the largest carbon sinks on the planet, absorbing it from the atmosphere; conversely, disturbing the soil releases carbon and CO2, one of the most potent greenhouse gases. We see the same effect from the felling of trees, which also are important carbon storehouses.

The Green Revolution's drive to produce ever-larger quantities of food depended on higher levels of chemical fertilizer and pesticide use, increased reliance on irrigation, and a voracious expansion of cultivated land. These became the hallmarks of our modern agriculture system, along with a focus on a small number of crops that could be optimized to provide cheap calories for both humans and livestock. Monocropping—carpeting fields with the same single crop—became the norm. Farmers were encouraged to expand their productive land by uprooting natural habitats vital for stable ecosystems, cutting down forests, and filling in wetlands, in order to sow these predominant crops from fencepost to fencepost. Of the thousands of plant species cultivated for food throughout history, only three—corn, wheat, and rice—provide nearly 60 percent of daily calorie consumption on a global basis today.

In this push to increase harvests and end hunger, agriculture has unleashed forces that now threaten those very goals. As Abebe's family saw on their farm in Ethiopia, agriculture can turn on itself, intensifying the already difficult challenges of farming. Agriculture's greatest allies—the soils that cradle the seeds; the trees that shade and feed the crops with nutrients and aid the precipitation cycles; the biodiversity of flora and fauna that enriches the environment—lie badly wounded.

The growing reliance on a handful of crops has made the global food system more vulnerable to pests and disease. It has sapped the soils of nutrients and diminished environmental biodiversity. It has relegated thousands of indigenous crops, which played a vital role in nourishing populations for generations, to museum relics, buried deep in doomsday seed vaults to guard against total extinction. The push to increase yields depleted vital water systems. Thicker application of fertilizer and pesticides introduced new pollutants to our soils, waters, and air. Deforestation to clear more land for crops and cattle has squeezed wildlife habitats and led to an unprecedented collapse of pollinator colonies—butterflies, bees, midges—threatening the germination of many plant-based foods.

The World Wildlife Fund estimates that 75 percent of the earth's ice-free land surface has been significantly altered by changing land-use dynamics, damaging diverse ecosystems and habitats; its global Living Planet Index of 2020 calculated an average 68 percent decrease in population sizes of mammals, birds, amphibians, reptiles, and fish since 1970. The Global Footprint Network, an international sustainability research organization, estimates that humanity's demand for natural resources—mainly for food, clean water, and shelter—exceeds the capacity of the planet's ecosystems for regeneration by about 70 percent. It marks an annual Earth Overshoot Day, the date when we have used all the biological resources that the earth can renew during the entire year. In 2023, it was August 2.

The actions of our global food system—the release of carbon once stored in soil and trees, the emission of methane from belching livestock, the production of food packaging that fills our garbage dumps along with the billions of tons of food we throw away each year—are collectively responsible for about one-third of the world's greenhouse gas emissions. The food and agriculture sector rivals the energy and manufacturing sectors as the largest emitter of destructive pollutants that contribute to warming temperatures and changing climates. A 2023 report by the UN's International Panel on Climate Change (IPCC) warned that the world is likely to bump up against what scientists consider relatively safe levels of warming—about 1.5 degrees Celsius, or 2.7 degrees Fahrenheit, above pre-industrial temperatures—in the next decade. An analysis in the Carbon Brief reported that about half of the world's population is living in regions that in the past decade saw their hottest daily temperatures since 1950. In 2022 alone, it calculated, about 380 million people saw their hottest single-hour temperature ever recorded. Then, 2023 became Earth's hottest calendar year in global temperature data records going back to 1850, according to the European Union's Copernicus Climate Change Service; it noted that each month of 2023 from June to December was warmer than the corresponding month in any previous year.

Efforts to hold off and reverse this trend toward warming temperatures have largely focused on the need for a dramatic reduction in the use of fossil fuels. But it is clear that a transformation of our modern-day food system is also imperative—most immediately, for agriculture's own sake. For those greenhouse gases released by agriculture activities are, paradoxically, most profoundly impacting agricultural production. The increasing tempo of extreme weather events around the world—prolonged droughts, intense flooding, scorching heat waves—stems from changes in climate driven in part by those warming gases. These weather events are wreaking havoc on harvests, weakening crop nutrients, altering the geography of farming, and forcing farmers in some regions to uproot their families and relocate, if not give up entirely. The majority of the world's climate refugees are farmers. In East Africa, a severe drought that stretched from 2020 into 2023, the longest in about seventy years of reliable rainfall records, ruined crops, killed livestock, and pushed millions of people to the brink of famine; by the end of 2022, the World Food Program estimated, about twenty-three million East Africans were "severely food insecure," nearly one million children suffered from acute malnutrition, and another million people were on the move in search of food and water. On the American Great Plains, a prolonged drought in 2022 and 2023 reduced wheat yields in some areas to their lowest level in a century. On the Indian subcontinent, a heat wave in 2022 that pushed temperatures above 120 degrees Fahrenheit scorched crops and reduced harvests by up to one-third of usual yields.

All of this has made the humanitarian aspiration of the Green Revolution more elusive than ever. Hunger and malnutrition continue to threaten us. Incredibly, in the third decade of the twenty-first century, global hunger was again on the rise in the era of the catastrophic Cs—Covid, climate, conflict. The coronavirus pandemic, a series of extreme weather outbursts, and the war in the grain belt of Ukraine all combined to severely rattle the global food chain, igniting a series of acute hunger emergencies. No region, no country was spared. In the United States, which considers itself the world's breadbasket, food lines formed at lengths and to a degree of desperation not seen since the Great Depression. In Africa, nearly three hundred million people—more than one-fifth of its population—were impacted by hunger in 2021–22, according to the International Fund for Agricultural Development, and one-third of the continent's children under five years of age were malnourished. Globally, nearly 25 percent of the world's children under the age of five were stunted, either physically or cognitively or both, from early childhood malnutrition, and more than 7 percent suffered from wasting.

Collectively, these shameful numbers mock the commitment of the global community to Goal 2 of the UN's Sustainable Development Goals, adopted in 2015: to achieve a world without hunger by 2030.

Even before this eruption of food emergencies at the beginning of the 2020s, the number of chronically hungry in the world—those lacking enough food from day to day, no matter whether an acute food emergency was flaring—stubbornly remained above three-quarters of a billion people. And more than twice that many, though perhaps getting enough calories daily, were micronutrient deficient, suffering a lack of crucial vitamins and minerals known as "hidden hunger." The modern food system's focus on calories and ultra-processed products have relegated nutrition to an afterthought. Another two billion people are overweight or obese, and that number is rising, as is the incidence of diet-related noncommunicable diseases like diabetes and hypertension. Poor diets are now a leading cause of deaths globally. As the world's food production has increased, both the planet and its people have become sicker.

◈

Norman Borlaug feared the fulfillment of his most haunting prophecy. In his Nobel Peace Prize acceptance speech in 1970, he warned against complacency. If hunger persists while food becomes more abundant, he said, "We will be guilty of criminal negligence, without extenuation. . . . Humanity cannot tolerate that guilt."

In his final years, while in his nineties, he railed against the complacency that indeed set in, even as the consequences of the Green Revolution and the resulting modern agriculture system became clear. He and many of his disciples began to acknowledge that unabated chemical fertilizer and pesticide use had created a new form of pollution, that unchecked irrigation had lowered water levels, that regimented cropping systems were impoverishing soils and diminishing nutritional diversity. He saw that all this was undermining the very goals—growing more food, ending hunger—he had set out to accomplish. His agricultural ally in India, M. S. Swaminathan, insisted that the Green Revolution wasn't meant to be a static moment in time but rather an Evergreen Revolution—an *Ev-olution*—that needed constant vigilance, adaptation, moderation, disruption. By neglecting the negative consequences, it seemed that agriculture's practitioners, investors, and funders had lost common sense.

Above all, Borlaug lamented that the interests of the world's five hundred million smallholder farmers—those trying to feed their families and communities on just a few acres of land, the ones who had been at the forefront of his efforts, the ones who grow one-third of the planet's food and own or manage nearly half of the world's forests and farm landscapes, the ones who themselves are the poorest and hungriest people on earth—were being steamrolled by the conformity of the modern agriculture system.

They were meant to be the beneficiaries of the Green Revolution, but were now the ones being buffeted the most by its consequences. These smallholder farmers, working mainly with their hands in fields watered only by rain, dependent on the kindness of the climate, are the least equipped to adapt to environmental changes.

Even in his dying days, in 2009, Norman Borlaug's thoughts remained with them. "Take it to the farmers," he told family members and close friends who gathered around him. They understood it as an exhortation to take the science and technology to the smallholders' fields, as he always intended. But also to take account of the farmers' experiences, wisdom, and questions.

Like the simple, revolutionary question Abebe Moliso asked: "Now why would I do that?"

After the Green Revolution, we thought we had all the right answers. It turns out we have all too often been asking the wrong questions. Or asking no questions at all.

◈

Another tall Ethiopian strides forward, this time from the fields of science and academia, to challenge us with his own set of urgent questions.

Gebisa Ejeta was born and raised in a rural village, like Abebe, and grew up to deliver millions from hunger as the planet's foremost sorghum breeder, which earned him the 2009 World Food Prize. Now, from his Center for Global Food Security at Purdue University, in the heart of America's Midwest breadbasket, Gebisa summons the ambition of a fellow Purdue grad who was the first person to walk on the moon. "That's one small step for man, one giant leap for mankind," said Neil Armstrong as he stepped onto the lunar surface in 1969. Gebisa believes the global agriculture community, for the sake of all mankind, desperately needs to take its own giant leap. Not a moon shot, but an earth shot.

"Human society in the past has shown that it can achieve extraordinary feats and solve big societal problems when it builds sufficient common resolve and will," he reminds a gathering of scientists, conservationists, humanitarians, and farmers at his Purdue Center. "Can it now build one such global resolve as a last-ditch effort to eradicate hunger from the face of the earth? And save the planet at the same time?"

His vision of the way forward demands a revolution more

far-reaching than growing additional food. The task ahead, Gebisa argues, is not solely a matter of seeds and soils, but also of equity and justice. It requires fostering a universal push to redress the unequal distribution of global wealth, resources, and knowledge—as well as forging a fresh era of human cooperation that unites a shared concern for agriculture, nutrition, environment, climate, biodiversity, and the good health of humans, animals, plants, insects, and all living things on the global food chain. Success will only come with a greater appreciation and respect for nature, local ecologies, and the common good—and a great leap of faith, where farmers and conservationists are no longer foes but allies. It also demands heeding the experiences of smallholder farmers we have far too often ignored, marginalized, and belittled.

One more question stirs Gebisa—as it should us all: "If not now, when?"

◈

This is where the dirt road in the Ethiopian highlands leads, to where Gebisa's countryman Abebe is taking his own giant leap in seeking to both nourish his family and restore and preserve his land. In this endeavor, he joins a legion of other smallholder, Indigenous, and family farmers around the world, some of whom we will meet in the coming pages. Toiling in some of humanity's ancient agriculture zones—throughout the Great Rift Valley in Africa's Horn and its vast eastern savannas; on the Indo-Gangetic Plain sweeping beneath the Himalayas on the Indian subcontinent; on the volcanic highlands of Central America; on the Great Plains of the American Midwest—they may show the way for the rest of us.

If not them, who?

Far from the air-conditioned climate change conferences where contentious conversations have mainly focused on the impact of industry, energy, transportation, and cities, these farmers are on the real front lines of the degraded fields, depleted soils, dwindling water sources, and disappearing biodiversity, contending with the higher temperatures, longer droughts, and sudden floods of our ever more extreme weather. All around them, populations keep swelling and the cries for food keep rising.

While scientists hunt for big-game solutions like vaccines to reduce the methane emissions of livestock or gene-editing crops to resist pests and withstand extreme weather, the smallholder, Indigenous, and family farmers are pushing forward with practices called agroecology, agroforestry, and regenerative agriculture, and leading the movement toward food sovereignty. They are farming against the grain of modern industrial agriculture, challenging conventional wisdom and long-held articles of faith as they reconcile their roles as both food providers and stewards of the land.

The Great Collision of humanity's two supreme imperatives—nourish the planet and preserve the planet—is upon them. They have seen the future, experienced, lived it. And it is ugly. They will tell us that, if we listen.

And these farmers will also tell us that it isn't inevitable. Listen to Abebe. He knows, for his farm, once dead, lives again.

CHAPTER 1

THE GREAT RIFT VALLEY: ETHIOPIA

What Have I Done?

Some thirty million years ago, two tectonic plates in east Africa began pulling away from each other, creating the geographic masterpiece that became known as the Great Rift Valley. The geologic tumult blessed Ethiopia with breathtaking natural wonders: escarpments, plateaus, mountains, valleys, lakes, deserts, savannas, forests, and some of the most fertile agricultural land on the continent.

Thirty years ago, one of those treasured features, the Humbo forest in the great lakes region of south-central Ethiopia, had all but disappeared. A very different set of disturbances—this time man-made, due to the relentless clearing of trees to create more space for people and for crops and live-

Left: *Back Home*
The farmers of Wolaita gather in the shade of an acacia tree

stock to feed them—had wiped out the forest. Woodlands and forests are the earth's lungs, breathing life into the environment. They create oxygen, sequester carbon dioxide, regulate microclimates, stir the cycle of precipitation, cool with shade and the rustle of wind, and host a riot of biodiversity above and below ground. All of these are vital partners of agriculture. As the forest disappeared, leaving behind a barren landscape with a blast-furnace climate, so too did the food from the neighboring farms.

"Our misery started when our cattle starved and our crops diminished. You couldn't find a family that hadn't lost a child because of drought and hunger," remembers Yissac Nadamo, a farmer who grew up on the edge of the Humbo forest, near the town of Wolaita-Sodo. Adds his friend and fellow farmer Abdullah: "Even mourning was difficult. There were too many deaths. You couldn't even cry anymore. You realized you might be the next."

The drought and resulting famine of 1984–85 is estimated to have killed upwards of one million people across Ethiopia, and triggered Wolaita's wave of misery. Some families abandoned their unproductive land and moved in search of greener pastures. But most stayed, persisting in their efforts to eke out something from the depleting soil. Until 2003, when another severe drought caused another wave of deaths. Local authorities ordered everyone, all 170 families, to leave before they perished. It was their last chance at survival.

"I was sad and angry," says Abdullah, a teenager when his family moved off their ancestral farming land. "There was a lot of disappointment then. We did it to ourselves. It was our own fault we had to move."

With the people and farming gone, nature went to work. The regeneration began. As with Abebe's nearby land, ruination slowly turned to salvation.

Now, back on their land, nearly two decades later, Abdullah and Yissac gather with a dozen other farmers to share what they learned. The men wear traditional straw stovepipe hats with wide brims and the women wear light, colorful scarves as the midday sun bears down from a cloudless sky. They carry stools and chairs from their round huts and settle under the broad, flat canopy of a young acacia tree.

A round of tea is poured. One farmer shares a honeycomb dripping with nectar, freshly retrieved from her beehive. In the thorny branches above their heads, noisy weaver birds go about their business of fashioning their hanging basket-like nests. And the farmers begin telling the story of a transformation they describe as miraculous.

This acacia tree, Yissac says, was one of the first to flourish after the farmers left. "For so many years, we couldn't be sitting here. It would have been too hot," he notes. "Now we have shade. And a breeze! And the birds are back, and their nests. Isn't it nice?"

One by one, the farmers acknowledge the mistakes they made individually and collectively. Pressed by the immediacy of hunger, the swelling population, the traditions and demands of their agriculture practices, they acted selfishly. They ignored the needs of nature. They cut down trees for immediate use as fuel and construction materials. As old farming fields became less productive, as grasses disappeared, they cut down trees to clear more land to grow crops and graze cattle. Why preserve land for later? When you have nothing to eat today, who cares about tomorrow? They took their abundant environment for granted. And then it was all gone.

"There was nothing left taller than your knees," Abdullah remembers.

The soil hardened like pavement in the unimpeded scorching sun. When the rain came, it ran off in flash floods. The earth was unable to absorb it. The farmers' fields, which had benefited from the presence of the forest, became an ecological wasteland. "Farming was so difficult," Abdullah says. "Impossible," adds Yissac. "It was terrible."

It was no better at their new location after the forced removal. The transplanted farmers were crowding other families, imposing on their fields. How long before this new land was destroyed as well, they wondered? So, many of them decided to make day visits back to their old home and give nature an assist in the regeneration. Five hundred people pitched in, says one farmer. No, more like six hundred, says another. They formed a citizen watch around the vanished forest so no one would encroach on the new growth. With aid from government extension advisors and organizations like the World Food Program (WFP), they planted seedlings and shrubs. On their abandoned fields, they attacked the concrete-hard soil with their shovels and, like Abebe, shaped terraces and planted grasses and dug water pans. They celebrated when the rain stayed instead of quickly fleeing. Over the next dozen years, they watched life return. They celebrated the appearance of every butterfly and bee, and every flower that blossomed, as signs that the time was nearing when they themselves could permanently return. When they did, they found that the restored forest and renewed biodiversity rewarded them with more food and income opportunities than they had before. Beyond the acacia tree were rows of maize, beans, onions, cassava, cabbage, coffee, and sorghum, and emerging orchards of mango, papaya, and avocado trees.

"Look, over there, we have ducks now," Yissac proclaims, pointing to a small flock landing on one of the newly formed ponds.

A butterfly flutters past the gathering under the acacia tree. More rejoicing. I mention how butterfly sightings have become rare and precious in parts of the United States as well. How the monarch butterflies, staples of Midwest summers, have become scarcer as farmers have plowed under milkweed plants (favorites of the monarchs) to squeeze in a couple more rows of corn and soybeans. How other familiar species vanished from old habitats as wetlands were filled to create more space for crops. The farmers nod, connected to fellow farmers on the other side of the world. "We know how it goes," Yissac says.

He raises his tea and offers a prayer of thanksgiving that the butterflies and bees, and their livelihoods as farmers, have returned. "I was very happy when we could come back," he says. "It would be our even greater happiness if this can be replicated elsewhere."

◈

As humans, we have long looked to this region of the world, to Africa's Great Rift Valley running from the Red Sea south through the continent's vast eastern savannas, for clues about our past. That's what a team of anthropologists were doing in November 1974 as they walked across the ancient sands of the Afar Desert in Ethiopia. They were heading back to their vehicle when they spotted a bone fragment they identified as a hominid forearm. Soon they found a skull, femur, pelvis, ribs, and a lower jaw. Within two weeks of digging, they had recovered 40 percent of a single female skeleton, which they estimated to be about 3.2 million years old. They named their discovery Lucy, after a night of celebrating and dancing to the Beatles' song "Lucy in the Sky with Diamonds." Lucy, in turn, was celebrated as our oldest known ancestor.

Over the past century, the Great Rift Valley has yielded

many sites where fossils of early hominids have been found, including the discoveries by the Leakey family in Kenya and Tanzania, and the uncovering of another female skeleton, named Ardi, also in the Afar Desert in 2009. These discoveries confirmed Africa as the cradle of humankind, and the Great Rift Valley as the site of an important stage of our evolution. Lucy opened a window to our past, revealing that our human ancestors were walking upright earlier than had been thought. The study of Lucy helped form the theory that as the Great Rift widened, early hominids left the forests and ventured out into the open savanna. Walking upright freed their hands for the creation of tools, and the brain evolved over time with the exploration of new possibilities. From her current repose in the basement of the National Museum of Ethiopia, in Addis Ababa, Lucy remains a guide to our past.

Today, her descendants, like Abebe, Abdullah, Yissac, and the other farmers tilling the ancient soils of the Rift Valley, are our guides to the future. Here, instead of a rift in tectonic plates, the Great Collision is the seismic force now threatening humanity. The impacts of climate change on the farmers' agriculture are manifest in recent catastrophes of biblical proportions: drought, floods, hail, locusts and other pests, and pestilence such as maize lethal necrosis disease. The impacts of the old slash-and-burn farming on the environment remain. And at the same time, all around the farmers, the demands to feed a rapidly growing population are escalating.

When I first visited Ethiopia in 2003, it was the world's fourteenth most populous country, with more than 70 million people. Twenty years later, it was twelfth, with about 126 million. By 2050, Ethiopia is projected to be the ninth most populous country, with about 213 million people. By 2100, it is estimated to be the seventh most populous, with nearly 330 million. That is like adding two and a half million more people to the dinner table every year of the twenty-first century.

The neighboring countries in Africa's Great Rift Valley are also among the fastest growing in the world. Kenya's population has swelled from 30 million in 2000 to 55 million in 2023, with projections of 85 million in 2050. In the same time frame, Uganda is projected to expand from 24 million to 47 million to 87 million, and Tanzania from 38 million to 67 million to 129 million. By 2050, projections place each of these four countries among the top twenty-one most populous nations in the world.

These growth rates are similar throughout sub-Saharan Africa. In 2023, the region's population hit about 1.4 billion, roughly 17 percent of world population. By 2050, it is projected to nearly double to about 2.5 billion, which will be fully one-quarter of the people on the planet. Over that time, Africa is expected to account for more than half of the world's total population growth. All this on a continent that already has the highest food insecurity in the world. More than 20 percent of Africa's population is chronically hungry and at least one-third of the children are malnourished, according to the United Nations' food agencies.

This puts enormous pressure on Africa's smallholder farmers, who produce the majority of the continent's food by tilling just a few acres of land mainly by hand and are largely dependent on the increasingly mercurial rains to water their crops. Despite their efforts, Africa's food import requirements continue to rise; the continent imported more than one hundred million metric tons of cereals at an annual cost of about $75 billion, the African Development Bank reported in 2023.

Africa's smallholders toil on some of the world's most degraded landscapes and in some of the most depleted soils;

it is estimated that as much as 65 percent of productive land in Africa is degraded, with more than twenty-four billion tons of fertile soil disappearing each year due to erosion and expansion of deserts and about seven million acres of forest cover lost annually. And they face the greatest vulnerability to extreme weather events and other environmental shocks, with the fewest resources to adapt. In Ethiopia, for instance, about 85 percent of agricultural output comes from subsistence plots of less than five acres, according to the WFP. Because of severely degraded land and the impact of unpredictable rains and recurrent droughts, the productivity of Ethiopia's smallholders, in yield per acre, is among the lowest in sub-Saharan Africa.

The ability of Ethiopia's smallholder farmers and those across the Great Rift Valley to rehabilitate and restore the productivity of their degraded lands holds the key to Africa's future—and ours. We, and all living things on this planet, are jointly connected by the global food chain, which is only as strong as its weakest link. And many contend that for now, that weakest link is here, on the continent where humanity began.

◈

As the twenty-first century dawned, Ethiopia was in the grip of twin disasters. One disaster was humanitarian: a merciless drought coupled with failed agricultural markets had brought fourteen million people to the brink of imminent starvation. And one was environmental: the worsening land degradation in parts of the country had forced many millions more to confront the prospect of a perpetual hunger season. Here, in 2003, at this juncture of great suffering for people and planet, the Great Collision had arrived, a harbinger of what loomed for the entire planet.

At the time, I was traveling with the WFP, which bore the

Herculean task of feeding those fourteen million, orchestrating mass relief caravans of trucks filled to the brim with bags of wheat, corn, and beans and cans of cooking oil. We headed south from the capital, Addis Ababa, on a highway originally paved with the good intention and grand ambition (ultimately unfulfilled) of connecting Africa from Cairo in the north to the Cape of Good Hope at South Africa's tip. We drove through the market town of Nazareth, past the village of Mojo, through the Rastafarian enclave of Sashemene, beside the chain of Great Rift Valley lakes, and beyond the main southern city of Awassa. Just before reaching the famed coffee growing region of Sidamo—source of some of the tastiest brews favored by customers in the richer areas of the world—we made a right turn off the paved highway and onto a dirt road that corkscrewed up an escarpment to the Boricha plateau. We hurried across the table-top flatness, past drought-stricken crops, to a field where the normally bustling farmers' market had become a calamitous hunger triage zone. Here was the humanitarian disaster of the Great Collision.

A cluster of large canvas tents were each filled with dozens of starving children. It was a ghastly scene that should shame the world's conscience—such medieval suffering still, in the twenty-first century. Inside one of the tents, I met Tesfaye, a smallholder farmer. He was sitting forlornly on a thin mattress on the ground. He cradled his son Hagirso between his legs. Hagirso, about five years old, weighed just twenty-seven pounds when his father carried him into the tent.

As a doctor and nurse examined Hagirso, they told Tesfaye his child might not survive; although he was beginning to recover with the emergency feeding treatment, the hunger shock had been so great. Tesfaye asked, "What have I done to my son?" Here was a father, a farmer, tormented by his failure to feed his own child. Had his actions as a farmer, trying to squeeze something from his tired soil, led to this, his starving boy? I was speechless—how to answer *that*?

The medics continued with their rounds, and I followed. But my thoughts remained with Tesfaye and Hagirso. Gradually, over the years, the beginnings of an answer formed. Tesfaye's haunting question needed to be amended and redirected so that it could be posed to all of us. If we heard only the plaintive cry of the farmer—what have I done to my son?—then we got the question all wrong. The right question was then, and remains now, what have *we* done to his child? A collective "we" brought hunger and malnutrition into our new millennium.

It is, as the Great Collision unfolds, the question of our time.

Tesfaye's guilt emphasized our guilt—our neglect, our fault. Here, in these emergency feeding tents surrounded by the misery of smallholder farmers, was the dashed promise of the Green Revolution. It was, as Norman Borlaug had prophesized, a burden too great for humanity to bear. How had it all gone so wrong?

After leaving the emergency feeding tents, we continued our journey through the hunger hot spots, traveling a bit further south, veering off toward the vanishing forests of the great lakes region, including Humbo. Here was the environmental disaster of the Great Collision.

It was as if we were driving on the edge of a cliff. Deep gullies lined both sides of the paved road, the soil having washed away over the years. What soil remained was rock-hard; with very little vegetation, it looked like an endless field of reddish-brown bricks. The only clusters of trees seemed to be on the sacred ground surrounding Orthodox churches—"tree museums," the Ethiopians called them.

My fellow travelers from the WFP worried aloud that the

feeding tents would become permanent features of the landscape, the farmers and their families becoming dependent on food aid forever, if something wasn't done to heal the land. In fact, given the relentless population growth, which multiplied the pressure to grow more food, the problem would only worsen.

Even before the 2003 crisis, the Ethiopian government estimated that more than eight million people were chronically food insecure, most of them in farming households unable to grow enough for themselves, mainly because they lived on these severely degraded lands. They were perpetually in need of food aid. The drought of 2003 added another six million who became food insecure when their crops, planted on marginally arable soil, failed that season. The solution, the WFP and the government knew, wasn't in bringing in more food from around the world—year after year after year—but in bringing the land back to life so the farmers could feed themselves and others.

The world's largest humanitarian organization would need to become a major environmental organization as well. And so, as it fed the fourteen million, the WFP began a new program called Managing Environmental Resources to Enable Transition to More Sustainable Livelihoods, or MERET. This initiative would bring together pieces from various practices that had been launched since the famines in the 1980s. Those early efforts were largely top-down, steered by the federal government, with little local community input and coordination. MERET would center its work in the communities; it would be the program that would support Abebe Moliso's family, and the farmers of Wolaita, as they restored their lands. The idea was to create a productive safety net program where food aid would become payments to families putting aside their farming and working instead on land rehabilitation, rather than a passive safety net of food aid handouts.

"After continuously cultivating the land for generations, it needs to lie fallow, unused for several years," explained Adane Dinku, who was the MERET coordinator at the time.

Adane was leading us to an initial plot to be rehabbed under MERET, several hundred acres on the edge of the vanishing Humbo forest. As we walked, he outlined the soil and water conservation tasks: closing off the area to agriculture and reshaping the land with terracing, gully rehabilitation, stone barriers, and reforestation to hold the soil; water harvesting such as pond construction, shallow well digging, and natural spring resurrection; managing soil fertility by composting with manure, ending the monocropping of grains with more diverse varieties including vegetables, fruits, and cattle fodder, and introducing new crops that would add nutrients like nitrogen to the soil.

"In our obsession for production," Adane said, "we forgot about conservation."

We arrived at one of the first stone enclosures to be constructed. "Progress!" Adane exclaimed, as he stepped around a puddle of water. It had rained the day before; rather than running off through the gullies, the water and soil was kept in place by the stones. There was mud, glorious mud. It was the first promising sight in a day that had presented only misery.

The inspectors bent down to admire the puddle and the mud. "This water is a precious thing," said Erkeno Wossoro, the MERET project representative from the Ministry of Agriculture. Volli Carucci, a WFP water specialist, scooped up the rainwater with his hands and savored the feeling as it dripped through his fingers. "It's gold," he said.

Staring into the puddle, they could see the future. "If we can preserve the water, we won't ever need to plant grasses. They will just grow," Adane said. He envisioned future rains soaking into the soil, raising the groundwater levels and feeding springs and ponds, filling wells to quench the thirst of people and livestock and to irrigate the crops.

"In a few years, we can start putting in vegetables, all sorts of crops, chilies, potatoes, beans, and maize," he said.

Added Erkeno, "As the new grasses grow, the cattle can return, we'll control the grazing. The farmers will even have grass to sell."

Erkeno said his country had no alternative than to reclaim the land. "In the Rift Valley, we see very degraded land, so much erosion. In developing countries water is so precious, soil is so precious," he explained. "Population is increasing and we are losing the soil. The highest food insecurity is in the highest areas of erosion. People don't understand that!"

With rising impatience, he asked a series of existential questions: How can we feed our people without conservation of this land and soil? We will have no future without soil. Without soil, where do you put the seeds, the fertilizer? How do you grow food?

Erkeno fortified his argument with Ministry of Agriculture statistics. He said about one hundred thousand acres of land were lost to cultivation because of erosion every year. Ethiopian smallholder farmers reaped an average of about one-half metric ton of maize per acre. So those acres lost to erosion every year meant about fifty thousand tons of lost maize production annually. Add that annual loss, he said, to the five million acres of land having already been degraded and another thirty million acres, with less than fifty centimeters of soil depth left, on the verge of being lost to cultivation. The bottom line of all that math, he said, added up to looming catastrophe if something wasn't done.

"If you don't pay attention to the land," Erkeno warned in 2003, "you are erasing the people. You are erasing the future."

◈

When I returned to this area of the Great Rift Valley nearly twenty years later, the people were still there, back on their farms, in ever greater numbers. Significantly, the farmers who had been relocated, like Abebe Moliso, Abdullah, Yissac, and others in Wolaita, had returned to their rehabilitated land and were growing food with more success than ever.

Adane and others from the World Food Program joined me for another walk over land MERET had set aside. As Adane had predicted two decades earlier, the grasses were indeed growing by themselves now. Thick, luscious grasses, waist-high in some places. The barren red-brown, brick-hard fields of 2003 were now soft green meadows, with shrubs and young trees and butterflies.

"Be careful," he cautions. "You can't see it because the grass is so tall, but the land here is terraced. That's how we caught the rains. It worked!"

As he walked, he reviewed the history. When this land was closed off from agriculture, there was virtually no vegetation cover. By 2017, when the MERET project was handed over to the local communities, the vegetation cover was 90 percent. Overall, MERET rehabilitated 168 sites with a total area of nearly 120,000 hectares, benefiting about 38,000 households. Eighty-four tree and plant nurseries had been established. Adane noted that an independent study found that nearly two-thirds of all MERET households perceived that they had successfully escaped from poverty once they were back farming again and were able to increase their income from harvesting surplus crops. A nutritional analysis concluded that more MERET households were consuming "acceptable" diets, thanks to greater crop diversity, as opposed to the "poor" diets of the majority of households that weren't part of the land rehabilitation program.

After thirty minutes of walking, we come across a spring. A few years after the grasses returned and the acacia trees began spreading their branches to shape their iconic flat-topped canopies, this spring suddenly appeared out of nowhere, propelled by a replenished and rejuvenated underground aquifer. The villagers celebrated, singing and dancing, as if it was oil gushing out of the ground.

The community pooled resources to harness the water with pipes and a pump so it would be more available to the community. They built a concrete basin around the spring, which had become a popular gathering place. They added a tank to store the water, and a tap where people could fill their buckets with water for drinking, cleaning, and cooking. They made sure to turn off the tap after every use, so that not a single drop would be wasted. The spring also fed a nearby river, which provided 160 households with small-scale irrigation on their fields.

The spring is a monument to all they have accomplished and sacrificed, and everything they lost. "So many people perished during the great famine," recalls Abraham Basa, leader of the local farming cooperative union. "Yes, some of my cousins and uncles died. Our cattle died. We had to do *something*."

They embraced the suggestions of the WFP and the MERET methods. The community discussed the priorities. First, they voted to seal off the farming fields and post guards to warn off intruders. They dug terraces and pans to collect the water in the fields. They let the grass grow and planted seedlings obtained for free from the nurseries. A collective decision was made to begin horticulture projects like growing vegetables and selecting drought-resistant varieties and crops, such as pigeon peas, that feed both humans and livestock while adding nitrogen to the soil. They selected a range of fruit trees: avocado, mango, and banana. From the sales of their surplus fruits and vegetables and bundles of grass, the community built a primary school and a clinic.

Members of the farmers' cooperative gathered on a grassy patch beneath a grove of eucalyptus trees they had planted. "Our fathers, grandfathers, great-grandfathers, they all came and took from the land," Abraham says. "They over-grazed and over-farmed. There was a cumulative effect of improper land use." This generation, he adds, honored the land. "We feel like we have brought back our land to its original status. Having experienced the worst of the drought and dying, we have done well to bring back the land."

The farmers note that neighboring communities come by to learn about forestry and conservation and collaborating with nature. "We have become a center of excellence," Abraham boasts.

◈

So too had a group of villages bordering the Humbo forest. Together, in 2006, they formed a regreening project led by the faith-based humanitarian organization World Vision. I joined several farmers as they scrambled up an escarpment that overlooked the forest and a series of majestic hills. Here was a grand panoramic view of the choices confronting the world today. Mostly it is green as far as the eye can see, a glimpse of a restored planet. But here and there are patches of brown that have yet to be rehabilitated. Several years ago, the farmers say, it was all brown.

Up until the famine of 2003, there had been open, unrestricted access to the forest. The swelling numbers of people took as they pleased: trees were felled for firewood, charcoal, and lumber; vegetation was mowed down by incessant cattle grazing; land was chewed up by farmers expanding agricultural fields. There seemed to be plenty of forest available—until there was none left. It was a tragic progression familiar to Abebe Moliso, to Abdullah and Yissac and the farmers of Wolaita, and to Abraham Basa and the members of his cooperative.

World Vision, an ally of the World Food Program in food aid distribution, was facing an endless relief effort in the Humbo area in 2003. An exit strategy through food aid, where they could declare "mission accomplished" with an end to childhood malnutrition, was beyond comprehension as long as the forest, and the farming it supported, continued to die. The organization was spearheading a movement that was gaining traction in other areas of Africa, particularly on the southern hem of the Sahara, to hold back the expanding desert. The movement was called "farmer managed natural regeneration," in which communities increased agricultural production through improved tree coverage. In the program, farmers protect and manage the growth of trees and shrubs, encouraging natural regeneration from stumps, underground root systems, and seeds spread naturally by birds, animals, and the wind. The new growth of trees and vegetation strengthens the soil structure, captures rain, and holds back erosion, all to the benefit of agriculture crops.

By 2006, World Vision organized seven farmer cooperatives surrounding Humbo to set aside more than six thousand acres that had been carved out of the forest—land cleared of trees by the farmers to grow crops, which had since become less and less productive. The farmers would assist in the regeneration, guarding the enclosure and nurturing the new vegetation. The World Bank joined the project, promoting a new carbon credit initiative that would pay the farmers for the carbon captured from the atmosphere by the freshly planted grasses and trees.

The farmers, who had never heard of the World Bank or

the notion of being paid for carbon capture, originally greeted the plan with plenty of skepticism. How could the air—the oxygen and carbon—be for sale? Could the forest really be resurrected? How long would it all take? If the results weren't immediate, what would they do? Would they ever get the closed land back, would they be able to use it again? Would the World Bank own it?

World Vision explained the benefits of trees, how they aided the farmers in growing food. In community meetings, they discussed how carbon dioxide in the atmosphere acts like a blanket, holding in heat and creating a greenhouse effect that warms the planet. Trees and vegetation absorb carbon from the air and release oxygen through photosynthesis. When trees are cut, less carbon is captured; when forests are burned, the carbon stored in the trees is released into the air. The World Bank would reward them for their tree-growing, carbon-capturing work.

It took nearly four years for the trees to grow and earn carbon revenue, but once the income stream began it continued to flow, and the farmers were impressed. Over the first ten years, the World Bank paid about $900,000 to the communities. With that money, each of the seven cooperatives constructed a grain storage warehouse. Nine new grain mills were built. Solar energy projects were launched. Individuals could access credit from the revenue to set up businesses.

The agriculture benefits were also clear: increased soil organic matter, decreased erosion and downstream flooding, fewer landslides on the hills, increased biodiversity with the return of wildlife and vegetation, and improved underground water levels from the rainwater soaking into the ground rather than running away. Within a decade, thirteen springs that were dried up had returned to life. New income opportunities arose, such as selling surplus crops and new products like grass for cattle fodder and honey from the proliferating beehives. The residents reported cooler temperatures, increased rainfall, and better air quality.

"To have our own grain mill now is such a relief," says Bogale Cholo, the chairman of one of the cooperatives. Before the mill, he adds, the farmers, both women and men, would carry their maize and sorghum harvests a dozen or so miles to the city of Sodo, site of the nearest mill that would turn their crop into flour. If they couldn't afford a donkey to haul the load, they would carry it on their own backs.

"Life was very hard," he says. "I remember, when I was ten, the forest was getting smaller and smaller. We would go into the forest whenever we wanted for firewood, cutting trees to make charcoal for cooking. It wasn't green like you see today. My parents were farmers. There wasn't always enough to eat. At times, a feeding center was set up here. There were food and clothes distributions. People were dying."

Now, instead of a feeding center handing out food aid, it is the grain mill that is at the center of community activity. With the return of the forest, food is abundant, he notes. During the harvest seasons, the mill grinds 700 kilograms a day. Farmers pay a few cents per kilogram for the service. "It's not much for the farmers, but it adds up for the cooperative," Bogale says. "That's what we use to finance the community improvements."

It also financed farmer Ata Sanadela's small loan, and his large ambition. He used the loan to buy a bull, which he put to work plowing his land and, for a fee, his neighbors' fields. With his harvest of fodder grass, a new crop since the return of the forest, he fattened the ox and then sold it, using the profit to buy other livestock. When his farming was so meager on the degraded land, his family depended on food aid. He now has enough income from his crops to pay the school fees for his children—and for himself. At the age of forty-five, he resumed his education, returning to the tenth grade. "I don't want to just be a farmer all my life," he says. "I have a dream to get a job and earn an income."

The restored forest has also fueled the dreams of farmers Soloman Satah and Wudinesh Masa. With his microloan from the carbon credit fund, Soloman bought a yellow sewing machine, powered by a foot pedal. He is the only tailor for miles around. Wudinesh, his wife, now stocks a small kiosk next door to the sewing parlor, selling snacks and daily essentials. While Soloman repairs a torn shirt, a snake is spotted outside his open door. "It's deadly," someone shouts. A crowd instantly gathers. Machetes flash, stones fly, and the snake is quickly dispatched. Soloman doesn't miss a stitch.

The Humbo rehabilitation progress has become a model for similar efforts throughout Ethiopia. Farmer managed natural regeneration had restored thousands of acres in the northern province of Tigray, which was a center of the 1984–85 famine. The land regeneration movement has also turned deserts green in Niger and other parched countries in West Africa. Tony Rinaudo, a World Vision agricultural specialist who has carried the program to many parts of Africa and beyond over the past four decades, marvels at the power of trees, grasses, and soils to heal the scars created by our efforts to nourish ourselves. "If you get the land issues right and help empower communities to overcome the impacts of climate change and degradation," he says, "you go from a vicious cycle to a virtuous cycle."

To keep the virtuous cycle spinning, a hub of tree and soil scientists are matching the wisdom and practices of smallholder farmers with cutting edge innovations in the labs of the World Agroforestry Center in the neighboring Great Rift Valley country of Kenya. In the Living Soils Lab, scientists with handheld scanners use infrared light and X-rays to analyze the health of soils taken from around the continent. They wave the scanners over dozens of petri dishes filled with soil in a range of colors, from maroon to dark brown to black. They can pinpoint mineral and nutrient deficiencies and traces of heavy metals and toxic material. They can also determine the impact of monocropping and tree-cutting on the declining health of soils, and then design mixtures of tree species and crops that can improve farm productivity. It's the equivalent of public-health surveillance for soils.

Another group studies the interactions of trees and their natural allies in the soil, the worms, centipedes, millipedes, ants, beetles, and termites. "Where there is a tree, a leaf drops and becomes litter that the insects and organisms break down into compost and carbon stores," one of the scientists explains.

Down the hall in the Food Tree Project lab, tables and shelves are covered with an array of fruits, nuts, and seeds that can be cultivated by farmers and communities to fill in seasonal hunger and micronutrient gaps. A tree gene bank preserves the seeds of hundreds of species, including indigenous varieties like baobab and moringa, that are gaining wider attention as medicines and superfoods.

The mantra of these labs is: the right tree for the right place for the right purpose and the right climate. "Agroforestry is the future of feeding people, just as wind and solar are the future of green energy," says Dennis Garrity, a former director general of the World Agroforestry Center. He was studying in the Indian state of Punjab in the early 1970s when the wheat and rice of the Green Revolution was sweeping through the region. "Now, we don't need to increase agricultural productivity all that much, but we need to transform agriculture using principles of ecology, to restore land, reduce emissions, accelerate tree cover on farmland, capture more carbon dioxide," Dennis says. He adds that returning more than a billion acres of degraded land to agricultural productivity would be truly revolutionary.

The first volleys were being fired by smallholder farmers in the Great Rift Valley and neighboring areas of Africa. A host of initiatives with ambitious titles like the Great Green Wall of Africa and Regreening Africa were taking root during the past decade in a number of countries, particularly in the Sahel region bordering the Sahara and the drylands of southern Africa; in both regions, a mighty battle was underway to hold back the advance of deserts. The stakes are high: land degradation and desertification cost the continent about 3 percent of its gross domestic product annually in squandered farming revenue. "Through desperation comes regeneration," Dennis says.

Led by the African Union, the Great Green Wall initiative aimed to nurture a five-thousand-mile-long corridor of landscape restoration across the widest stretch of the continent, from Senegal on the Atlantic Ocean in the west to Ethiopia and the Red Sea in the east. It pursued a similar restoration ambition in the degraded lands around the Namib and Kalahari deserts on the continent's southern tip. The African Forest Landscape Restoration Initiative set out to restore about 250 million acres of degraded forest land across the continent by 2030. And Regreening Africa targeted the restoration of two and a half million acres for the benefit of five hundred thousand households in eight specific countries by integrating trees, crops, and livestock (the practice known as agroforestry) and following farmer managed natural regeneration practices.

Ethiopia and the Great Rift Valley were at the center of all these initiatives, with smallholder farmers leading the way. In

July 2019, farmers and urban gardeners throughout Ethiopia set a world record for tree planting by a single country in a single day. At least Melese Mena Gemeso, deputy manager of trade and natural resources in the southern region, claimed it was a world record. "Three hundred million trees, in just one day!" he boasted. "My region alone planted fifty-five million!" The old record, he maintained, was held by India, with sixty-six million trees. That one-day effort was part of Ethiopia's plan to plant 1.6 billion seedlings every year. To achieve that, rural communities were asking residents to contribute one month of free labor to regenerate the land. "We are on the attack," Melese Mena Gemeso said.

Abebe Moliso and his neighbors, and Abdullah, Yissac, and the men and women of Wolaita, and the cooperative leaders Abraham Basa and Bogale Cholo and their members—they all planted.

And so did Tesfaye and Hagirso, the father and son I first met in the emergency feeding tent in 2003, at the height of the humanitarian and environmental crisis in the Great Rift Valley. Nearly two decades later, the land restoration movement had also arrived on their Boricha plateau. A peaceful green meadow now covered much of the field where the emergency feeding tents once stood. The small clinic expanded to a full-fledged hospital. In this place haunted by great suffering, new mothers and mothers-to-be gather on the grass of the hospital courtyard for a nutrition class. The nurses and doctors now practice prevention instead of triage.

At the nutrition class, local health officials and nutritionists from the World Food Program display a series of posters to demonstrate the importance of good nutrition during the "first thousand days"—the time from when a mother becomes pregnant to the second birthday of her child. They tell the mothers that if their children receive the proper vitamins and minerals in their food, they will do well in school and get a good education. And then they begin cooking. Using common ingredients—potatoes, sweet potatoes, squash, beans, greens, eggs—they prepare a nutritious vegetable-and-egg porridge and share it with the moms in a communal tasting session. Finally, the instructors measure the mid-upper arm circumference of the mothers and children. It is a measurement that allows health workers to quickly determine if a patient is acutely malnourished and in need of therapeutic treatment. The regular measurements are recorded in a notebook and followed over time, providing an early warning should anyone slide into malnourishment. "This is very important to keep us all healthy and strong," the nutritionist says. "We don't ever want to see malnutrition here again."

At Tesfaye's small farm, the recovery from the great drought of 2003 continued. Hagirso, his son who had been near starvation, did indeed survive. He was now a young adult, in his early twenties. His recovery, too, continued. The childhood stunting brought on by the severe malnutrition followed him through the years. Physically, he is thin and shorter than his father, and his immune system is weak; he is often ill. The greatest impact, though, has been on his cognitive development, setting back his school attendance. Tesfaye worried his son wouldn't be able to keep up with the lessons. Hagirso started school late, entering first grade when he was fifteen. Six years later, he was in a fourth-grade classroom, learning simple math. More than sixty other students crowded into the room, sharing desks and standing against the mud-packed walls. More than half of them were eighteen or older. Some began school late because their parents couldn't afford the school fees; others were kept home to help with the farm work. But most of them, like Hagirso, were behind in their education because of childhood hunger. Here was the lasting impact of the famine of 2003, a scar in the community that no landscape restoration can cover: a generation of children severely affected by malnourishment. "We have many Hagirsos here," says Yonas Markos, the school's principal.

The challenges caused by early malnutrition continue to follow Hagirso. His ambition is to be a businessman or a teacher, or a doctor or nurse—"To help people, like they helped me," he says—though he admits he is still struggling to master the alphabet and multiplication. As Hagirso returns home from school, he is eagerly greeted by his littlest brother, who is about the same age Hagirso was when his father carried him to the emergency feeding tent. Tesfaye calls him "Enough." When I ask why, Tesfaye laughs. "Because we said he will be our last child. God has blessed us with enough," he says. "And we have had enough of hunger and malnutrition. I pray we will have enough to feed him." But, he adds, "I am happy to say our situation is better than when we first met, in the tent."

Tesfaye says Hagirso is becoming a capable farmer, helping tend the maize, cassava, potatoes, beans, and greens, as well as their one cow and calf. Together, father and son have learned the importance of working with nature. They have diversified the family's one-acre plot, integrating more trees—avocado, mandarin orange, coffee, and banana—to help shade the crops beneath and to hold the soil. Both Tesfaye and Hagirso are involved in the monthly tree planting and land restoration work of the community, doing their part to hold back the Great Collision. They shape terraces and dig rainwater harvesting ponds. Hagirso says he joins the community youth brigade two days a week after school. "We dig and bring up the soil," he explains, "so the water can come and rest here and not run away."

This is the fervent ambition, and hope, spreading throughout the Great Rift Valley. That the water stays and rests in the verdant fields, and the crops and trees and grasses grow. And no farmer will ever again ask, What have I done to my child?

CHAPTER 2

THE GREAT RIFT VALLEY: UGANDA

Why Didn't We See This Coming?

ON THE WESTERN EDGE of the Great Rift Valley, where the Nile River emerges from Lake Victoria and begins its journey north through Uganda, farmer Daniel Sebakakyi is also setting out to relieve the remorse that haunts him. The sun has been up for only an hour, but already it is turning up the heat on the river town of Namasagali. Daniel feels partly responsible for this, and for the drying wetlands and the elusive rains of late. His role in it all began when he cut down his first tree to plant a new crop.

"I cleared my forest," he says. "I really regret what I did. And now I need to make it right."

Daniel had been growing maize, cassava, and peanuts on eight acres of fertile soil flanking the Nile in central Uganda.

Left: *Just Getting Started*
Jane Sabbi with the homegrown ingredients for her life-saving porridge

The crops were scattered amid forests and wetlands that enrich the local ecosystem. It was an integrated agriculture method, farming in harmony with nature, a practice that was keeping his family fed. But Daniel thought he could do better. He hoped to squeeze more income from his land to pay for the education of his four young children in the coming years. Sugar cane factories were sprouting up throughout the Nile basin, offering high prices for farmers who would sign up to grow for them. Daniel eagerly joined the trend. By his calculations, an investment of 14 million Ugandan shillings (about $5,000 at the time) to convert about five acres into sugar cane would turn into a harvest worth about $8,000, year after year. After the first cutting, cane continues to grow, like grass, providing steady yields over time. That was the plan.

"I thought sugar cane was more promising," he says. "I was convinced by those in the business. I saw what they were doing and bought in."

The thing with sugar cane, Daniel learned, is that it doesn't do well with trees. "It doesn't like the shade," he explains. So growing cane in a forested area meant clearing the trees, a lot of them. How many? "Probably a million, trees and shrubs," he says with only slight exaggeration. "It was a thicket. Everything had to be uprooted."

Daniel wielded a chain saw and joined the loud chorus of buzzing that became the background noise of Namasagali and across the wetlands. Everybody, it seemed, was cutting down trees to get in on the sugar cane rush. As the trees fell, the air was filled with the smoke of smoldering embers turning the logs into charcoal for cooking fuel.

Large termite mounds, some five or six feet tall, were also being leveled across the landscape. Maize can grow around and on top of them, taking advantage of the termites' soil-enriching activity. But sugar cane doesn't do well with termites, either. So Daniel poured pesticide down the mounds. The termites, along with the trees, vanished.

He planted the sugar cane and waited. The variety of cane grown in central Uganda generally takes about two years to mature before being ready for harvest. Daniel watched other farmers—more and more of them, he noticed—send their cane cuttings off to the factory. Long trucks hauling the cane sped up and down the roads like huge ants hurrying home.

As the time of his harvest neared, troubling news spread along the Nile: the sugar factory abruptly announced that it couldn't take any more cane, including Daniel's crop. It was running at full capacity. Daniel scrambled to find a new buyer, perhaps in neighboring Kenya, but he was unable to get a permit to take his cane across the border and sell it there. The sugar glut bred anger and jealousy among the farmers, leading to acts of sabotage. One night, with his lush green crop languishing in the field, Daniel received an alarming call.

"Your cane field is on fire," he was told.

He rushed to his field and found the fire still blazing, consuming his sugar cane, his hoped-for income, his ambitions. "I lost it all," he groans. He didn't even have one harvest. "It was all in vain," he says with a deep sigh.

He mourns the lost money. But he acknowledges an even steeper price was paid: Uganda lost an ecosystem as forests fell under the advance of sugar cane. "What have I done?" he asked himself, echoing Tesfaye's anguish. Without the trees to assist the precipitation cycle, rains became less predictable. Temperatures rose and everyone complained about the heat. Farmers, compelled by the pressure to feed and educate their children, turned from their dashed sugar cane dreams to expanding their cultivation of rice in the Nile wetlands. But as they did, paddy farming soaked up water that was an integral part of the environment and a vital need for other crops to grow. Rice is an especially thirsty crop. Through the decomposition of its organic matter in the water, it is also a leading emitter of methane, one of the greenhouse gases that trap the sun's heat and alter the earth's climate.

Daniel now recognizes that by his actions he has hastened the Great Collision in his part of the world. In fields all along the Nile, farmers' decisions and practices of the past are making present attempts to grow food and nourish their children that much more difficult. Maize, beans, cassava, sweet potatoes, and a vast array of fruits and vegetables are all suffering because of the altered environment and climate. Crop diseases become more lethal; new pests have appeared.

"The seasons have changed," Daniel says. "I see the clouds form and a little rain falls and then the wind comes and blows it away. The rainy periods are shorter. Farmers are late to plant because of late rains. There is too much sun for the crops to tolerate."

He torments himself with these questions: Why didn't we see this coming? Why didn't we know the consequences?

But he knows what he needs to do now. "I will replant the trees," he says.

He hops on a motorcycle and speeds off to his fields to demonstrate his vision. There he finds his youngest son tending to three cows and five calves. The ten-year-old boy is wearing a t-shirt from Mickey Gilley's saloon in Texas that made its way through the circuitous global used-clothes channels to Namasagali. His father says his four boys deserve a future with trees and a more stable climate.

The field where the sugar cane once stood still has a few reminders of its history. It is barren except for several lonely acacia and eucalyptus trees on the edges that Daniel left standing as markers of the past. "I'm showing you now the various species we had here," he says. "It was a natural forest." He sighs again. "So many trees I destroyed."

The word "regret" punctuates his sentences. Grasses and bushes are beginning to revegetate the land. Daniel's vision is to fill this field with a new eight-acre forest. He plans to select tree varieties that will grow fast and provide a canopy of shade for crops growing in between, and leave room for cattle grazing on the ground cover. He hopes the trees will one day provide income, particularly for school fees, as he culls the forest to make room for new growth.

Until then, for immediate income, he is cultivating rice on several acres in the wetlands portion of his farm. And watermelons, too. Both need rain, which has him looking to the sky for signs of clouds. "I regret the weather has changed," he says. He throws up his hands. "We're starving for rain."

He knows his rice takes a toll on the environment, and he promises that once the trees are growing and the forest is reestablished, he will get out of rice so the wetlands can heal as well.

But is it too late? Daniel desperately hopes that it isn't, that he still has time to redirect his agriculture to both nourish and preserve. At forty-five, he is thinking of his legacy. He wants the remaining years of his farming to be productive without also being counterproductive. "I'm growing older and older," he says. "We need to revive Uganda. We'll always need food. If we don't replant the trees, then it won't rain. And if it doesn't rain, we can't grow food. And if we can't grow food, it will all be in vain."

◈

A short drive farther down the dirt road, Salim Balyejusa also regrets his role in the changing climate. He once tended cattle, about fifty head, grazing in his forested land near the Nile. But it wasn't much of a livelihood; he struggled to feed and clothe and educate his seven children. So he turned to farming, intercropping with the trees and then planting in the meadows. First he sowed ten acres of maize, and then diversified on another ten acres, adding other local grains along with banana and cacao trees and passion fruit vines. "It was very productive and very nice looking," he says. The surrounding forest provided a protective frame and windbreak for his crops.

Then, in 2014, he made a fateful decision. Another farmer offered to rent a large portion of his land, dozens of acres, to plant sugar cane. Salim agreed, but soon came to regret the deal as the chain saws went to work clearing the forest. "Yes, I knew that would happen when I rented the land. It provided good income, but it has a side effect that is bad. It wasn't a good investment," he says.

The deforestation eliminated his tree cover. "After that, I could stand at a point up by my house and see the end of my land," he notes. "I couldn't do that before because of the trees."

Gradually, as the sugar cane fever spread to neighboring farms and more trees were felled, he saw the consequences worsen. "I tell you, it has changed the climate, brought more drought and extreme weather conditions," he insists. "It's true."

He recounts the changes. "We had prolonged droughts in 2017 and 2018, and again in 2022," he says. And then he describes the day when a fierce storm, with hail and strong winds, swept up along the Nile. Without the old-growth forest trees as a windbreak, the unimpeded gusts flattened his maize and uprooted his banana and cacao trees. The hail pelted his passion fruit.

One month later, he walks through the remnants. "It's all

destroyed," he says, stepping over the fallen fruit trees. More agriculture work, like Daniel's, done in vain.

Salim's words are drowned out by the loud buzzing of chain saws. A neighbor is clearing more forest. Salim cringes at the sound he has come to hate. "They don't realize what they are doing," he says. "They cut trees for sugar cane and charcoal and construction materials. They need the income. I tell them the changing weather is because the forests are being cut down. The trees are good for rain. They don't listen."

He too, like Daniel, is replanting trees. "Eucalyptus and cedar, along the boundary of my land," he says. "For every child, I plant ten trees."

It is a gift to the next generation, to reverse the damage of his. And he passes along a piece of hard-learned wisdom: "It's good to have a forest."

◈

When it comes to agriculture's impact on the environment, the spotlight often shines on the world's famous forests: the Amazon rainforest, under assault from the expansion of cattle grazing and soybean growing in Brazil and neighboring countries, and the tropical forests of Borneo and Sumatra, shrinking rapidly with the relentless advance of palm oil plantations, which also threaten precious orangutan habitat. But the deforestation is just as devastating in places where the spotlights rarely shine. Africa annually loses more than seven million acres of its forests, according to the African Forest Landscape Restoration Initiative. Uganda loses about 3 percent of its forests every year; the country's forest cover declined from nearly thirteen million acres in 1990 to fewer than five million acres in 2015. Uganda's wetlands, which cover an estimated 11 percent of the country, have declined by about 30 percent since the mid-1990s.

Those losses, caused largely by agriculture practices, rattle the entire food chain. The impacts reverberate socially, culturally, and personally.

"I remember, there was one tree that was used for the cooking fire. It had a smoke with a very sweet smell, a very different fragrance," says Gideon Nadiope, closing his eyes to savor a memory from his childhood growing up not far from Namasagali. "Now I can't find that tree anymore. I ask all over for it, has anyone seen it? Some younger people have never smelled that smoke."

His heartache over a time long gone is palpable. He recalls that an oft-heard saying in his hometown was "wait till the morning." Leopards prowled in the dense forest, making it too risky to pass through at night. "So we would wait till the morning." Now the leopards are gone from the thinned forest, as are the hippos that used to wallow in the wetlands, and the antelope, impala, and water buck that thrived in the mixed environment. During his youth, Gideon remembers, a horn would sound and the villagers would grab their spears for a day of hunting to restore their food stocks. These days, he notes, spotting wildlife is a precious excitement. The crested cranes that are Uganda's national symbol are an ever-rarer sighting in the nearby wetlands, which are the birds' natural breeding grounds.

Gideon grew up to become a veterinarian and the national director of the Center for Sustainable Rural Livelihoods established by Iowa State University (ISU) in the town of Kamuli, about an hour's drive north of the mouth of the Nile. In 2003, when the Center began its work, a shift in agriculture was picking up steam, a migration of farming from the dry upland fields, where harvests were dwindling and soil erosion accelerating, and into the forests and wetlands along the Nile.

"They were hoping for better yields. It was a coping strategy to preserve their livelihoods as farmers," Gideon says. One of the first things the farmers would do, he explains, was remove the papyrus and reeds, which held the water of the wetlands in place. Then they removed the trees. They were clearing more land to farm, but the environmental balance and biodiversity were never the same.

"The culture of trees is gone," Gideon says sorrowfully. "The practice of planting trees is gone, people think they just grow by themselves." He remembers planting a tree when he was about six years old. "An avocado tree. I really looked after that tree. It is still there, fifty years later now. I ask the people who live around there now, Where are the trees you planted? They had nothing to show. They haven't planted any. Instead, they have cut down many trees."

Gideon has made sure there are plenty of trees growing on the tidy campus of the Iowa State–Uganda program. It is a learning laboratory and, in effect, a tree museum, like the Orthodox churches in Ethiopia that provide sacred haven for trees. There are also demonstration plots for a wide array of crops, and enclosures for chickens, pigs, goats, and cattle, as well as beehives. From here the knowledge carried by professors and students from both Iowa and Uganda flows to the smallholder farmers along the Nile, intending to sustain their livelihoods without upsetting the forests and wetlands.

"This road used to be impassable during the rainy seasons. It would be difficult to travel at this time of the year," says Dorothy Masinde as she sets out to visit farmers. She is the associate director for the nutrition education initiative of the ISU–Uganda program. Now, instead of a mire of mud, the road carrying us to farms along the Nile is a bone-dry ribbon of brown dirt, baked in the sun to be concrete hard. Although it is the rainy season—at least by past weather patterns—there hasn't been a downpour for weeks.

"This used to be all swamp, a thicket of reeds and trees. You couldn't see much beyond the side of the road," Dorothy recalls. She has been traveling these roads for the past two decades, bringing farmers into the livelihoods and nutrition programs. "Now it's rice, maize, and sugar cane, with just a little stream. Farmers come into the swamp with ox plows, they dig it up and plant rice. Rice absorbs the water, then they come in with maize."

When this was all wetlands, she explains, picking up Gideon's narrative, the water would accumulate during the rains and move slowly toward the Nile. "When you grow rice, the water levels keep going down. The more rice, the less water flows. The area turns from wetland to regular land. But without water, you grow less rice. The farmers change the wetland to grow more food, but it also changes the environment, making it hard to grow food."

Agriculture turns against itself. As Daniel and Salim learned, farmers hoping to improve their livelihoods instead can endanger them.

Dorothy greets an old friend, Kitimbo Erieza Isabirye, who has been farming along the Nile for five decades. He is approaching eighty, a steward of the four hundred acres he has used over time to show the possibilities, and benefits, of both nourishing the planet and restoring it. Much of his land is covered by eucalyptus, acacia, and pine trees. "I'm trying to preserve the environment for future generations, for whoever has the energy," he says. "Look, you can see the difference here. Sometimes we get rain when other areas are dry. Yesterday, it rained some here. You can see how the road is still a little muddy. Until you arrived here at my place, I'm sure the road was dry and hard."

I ask if other farmers follow his lead. Does he warn them about the impact of cutting down trees?

"Do you think they listen?" he scoffs. "They don't. What's funny is they will steal your wood, they come and cut. You'll see them carry a eucalyptus down the road. I ask them, 'Where did you get that?' 'It fell down,' they say. 'What's the harm? It's just one.' That means it's nothing to them. Just a tree. But one cut tree leads to another and another. This is how we live today."

He drives his old Corolla past some patches of land he has set aside for farmers who don't have any, where they are growing maize and an assortment of vegetables to feed their families. "I haven't allowed anyone to grow rice or sugar cane to make other people rich," he says. "I have this garden land for poor people, desperate for something to eat."

Kitimbo arrives at his favorite spot, a meadow on a hill overlooking the Nile. He once had about three hundred head of cattle grazing here and in the forest. Now, as he's grown older, he's reduced the herd to about twenty cattle and nearly one hundred goats. Beside the pasture, he tends three acres of bananas, which he intercrops with cashew trees. "The cashew trees will grow big, like mango trees, and then the bananas will be under the cashews," he explains. "Bananas like shade."

He walks to the edge of the hill, where it slopes down toward the Nile. It is a beautiful, serene view. But Kitimbo is troubled about the future and how the farmers who come after him will fare. "If the weather doesn't change, it will be difficult for them," he says. "We used to have two rainy seasons in a year. Now it's so unpredictable. Rains come early, people aren't ready to plant. Then the rain disappears. Those that are able to plant early will harvest, those that are late will have nothing. Then they will cry 'hunger, hunger, hunger.'"

He wishes the government would act firmly to preserve

the wetlands and the forests. The official rhetoric at times sounds positive, he says, but then new permits are issued to allow farming projects in the wetlands or factories to carve out space in the forests. Kitimbo lists several places on the way to Kampala, Uganda's capital, where industrial developments have replaced forests.

One is right across the Nile from his cherished meadow. Forest land was flattened to make way for a sugar refinery. Cane grows all around it.

"I will never grow sugar cane. Never!" he says defiantly, pointing to the refinery. "This isn't my land, its God's land. I want to plant more trees."

◈

Down below on the Nile, in the small fishing village of Rubaizi, evidence of how the agricultural activities of the farmers have impacted the livelihoods of the fisher communities abounds. As the sugar cane and rice cultivation have eliminated forests and soaked up water in the wetlands, the natural flow of water into the river has changed. The water in the wetlands should carry nutrients for the fish from the forests and the grasses into the river. But when cutting the forests exposed more land to erosion, more chemical fertilizer runoff from the farmland ended up in the Nile.

The agricultural expansion that causes greater chemical runoff, combined with rapid urban growth around Lake Victoria, which increases human pollution in the lake and alters temperature levels of the water, have impacted incomes, nutrition, and fisher-farmer relations all along the Nile. Resulting algae blooms and the proliferation of invasive water hyacinths have sucked oxygen from Lake Victoria, which is the source of the White Nile branch that runs up through Uganda; the resulting eutrophication has triggered massive die-offs of tilapia and Nile perch, the larger fish that filled the nets of the fishing communities. Water diverted from a dam for industrial development flooded the Nile—water levels rose fifteen feet in the Rubaizi area—and submerged grassy areas near the banks that were favored habitats for the fish. The larger fish moved further upstream, replaced by smaller silver fish. The silver fish are high in protein, but many more are needed to feed a family and replace the income of the bigger fish.

"We don't have land, we fish. We don't know how to do anything else," says one of the fishermen, Moses. He and a dozen fellow fishermen toil in the shade of a yellow oleander tree at their boat landing, the Nile at their feet. They are untangling and mending their nets after a morning of fishing. A few dugout canoes still patrol the river. So far, no big fish in the day's catch. They worry, given their dramatically altered ecosystem, whether they can mend their livelihoods as well.

"Big industry controls the wetlands, with the sugar cane and rice," complains a second fisherman, leaning on his canoe paddle. Adds another: "You don't see much maize here, it's all sugar cane. Food is scarce and expensive. We just eat the silver fish, whatever is left over after we sell for income. It was better when we could eat the bigger fish. But they have moved away. Ei, ei, ei, how good things were."

Now, says Moses, "We're all so skinny here." It is a Monday morning. More children than fishers idle at the landing. "They should be in school. But we can't afford clothes or food for them. We can't send them to school in rags. And hungry."

He shrugs. "We have a saying, 'If we die, we die.'"

◈

When Dorothy first started visiting the farmers two decades ago, she heard a lot of talk about famine, not in general but in very personal ways. Mothers and fathers spoke of empty household cupboards and malnourished children dying from hunger. Some parents believed the malnutrition besetting their children was caused by witchcraft. They were selling the metal sheets of their roofs and even their pots and pans—any assets they had—to buy food.

"Ooo, that was a terrible time for us," remembers Johnson Mitala, one of the first farmers Dorothy approached. He says it got so bad that, yes, they even had no bananas, which was especially painful because Ugandans are known for their production and voracious consumption of bananas. "People from other places laughed at us. How can you not have bananas? You are in Uganda, and you need people to bring you bananas?"

There wasn't a particularly bad drought then, Johnson says. "It's just that we had such poor agricultural practices." There had been no agriculture development initiatives in their area, no extension advice from government officials. Soils were depleted. The yields of maize, the staple food in most households and the main crop grown by farmers, were miserly, maybe just two hundred kilograms from one acre—among the worst yields anywhere in the Rift Valley. His family, and so many others, endured prolonged hunger seasons, the period of dire deprivation between when the food from the previous harvest runs out and before the next harvest comes in. In the hunger season, food is rationed and the number of meals per day shrinks from three to two, to one, to none on some days. Farmers depended on food aid grown by other farmers in the world and distributed by the World Food Program. Now, he says proudly, his maize harvest has improved so much that he is *selling* to the WFP.

A bit further down the road, Paul Babi's house was falling apart. So when Dorothy arrived at his farm, that was her starting point. Iowa State students, along with Habitat for Humanity, helped to build a new house, and agriculture development experts helped him restore his farm. Paul, in turn, became a grateful and eager student, experimenting with new crops to diversify his fields and restore his soils. He was particularly keen to grow vegetables, and soon his seven acres, which had such poor maize yields, were covered in cabbage, tomatoes, collards, onions, soybeans, sweet potatoes, cassava, bananas, pumpkins, and squash.

It wasn't long before Paul became a trainer himself, walking from farm to farm, demonstrating how to work in harmony with nature. It was an important alternative, he told his neighbors, to farming in the wetlands and forests. "The wetlands might one day be banned for agricultural use by the government," he warned them. "So we have to learn to farm where we are, on the dryland."

In the Nile communities, Paul has become known as a pioneer of agroforestry, the practice of integrating trees and shrubs into crop and livestock farming. Instead of cutting trees, he advises, you work with them to improve your family's income and nutrition. "One, you chase hunger from your home. Two, you send your children to school. Three, you sustain the environment," Paul says. He preaches with an evangelist's fervor. But rather than speaking in tongues, he speaks in bullet points.

And he speaks a lot. Neighboring farmers often stop by, seeking advice. He enthusiastically welcomes the visitors and wants to host more. "My hope is that this farm will become a training center, so people can come and learn from me. One, they know I love the job of farming. Two, they like how I train them. Three, people hear about me by word of mouth." His vision includes building a gazebo to serve as a classroom for his lessons on agroforestry and the use of indigenous crops. If he builds it, he believes, they will come.

The trees and crops, he explains, will work together to improve the soil; one crop may pull a certain nutrient out of the soil (maize, for instance, depletes nitrogen) while another (legumes, like beans and the acacia trees) will put it back. Which makes mucuna beans a key addition to the crop rotation. "One, they fix nitrogen in the soil," Paul says. "Two, they are forage for livestock. Three, they are a cover crop, covering the soil, holding water."

Four, Dorothy adds, mucuna beans are high in protein and packed with nutrients; in richer parts of the world, they are sold as a superfood. Paul's farm is a garden of superfoods. He and Dorothy are sitting under a tall moringa tree, which is cherished for its many medicinal properties. "Tell them about the amaranth," Paul's wife suggests.

"Yes, the amaranth!" Paul says. Dismissed in the U.S. as "pigweed," amaranth, an ancient grain, is perhaps the most nutritious of Paul's plants, and a key ingredient of Dorothy's nutrition program. "You must see it," he says. "It is growing in the back field." He leaves the shade of the moringa tree and leads the way.

The walk to the amaranth winds through Paul's multistory agroforest masterpiece. On the ground floor, hugging the soil, are the vegetables and tubers: cabbage, collards, cassava, potatoes, sweet potatoes, onions, pumpkins, squash, mucuna beans, among others. "Oh, here's a nice red tomato," he notices while inspecting the plants. These ground floor crops provide soil cover. Paul has dug trenches to collect rainwater when it comes and direct it to the plants, and he added a borehole well from which he draws water that he sprinkles on the leaves. He has also constructed some raised beds to better conserve water; in the beds grow peppers, eggplant, and assorted beans.

On the second floor, rising above the vegetables, are banana, cacao, and coffee trees. The tendrils from the squash and pumpkins curl up the stems of these trees. The maize is on this level, too; after the cobs are harvested, the stalks are cut and added to the ground cover.

On the third floor are the fruits from tangerine, mango, papaya, and jackfruit trees, and passion fruit vines. And on the fourth floor, the penthouse, are the spreading canopies of the giant shade trees, the moringa and ficus and other native species—all providing valuable cooling shade. A local vine sprouting with what are called oyster nuts crawls up the ficus; the nuts are high in oil used for cooking and making soap.

Paul explains that each layer helps the other. The trees and plants all contribute nutrients to the soil, which in turn help them grow. As the leaves from the top fall, they provide ground cover and mulch for the vegetables. Every living thing has a purpose here. Paul admits it may all look chaotic and messy, with leaves and stalks lying all around, but it is all part of his agroforestry strategy.

His abiding principle: never cut down a tree without reason. "One, the oxygen we need comes from trees and photosynthesis. Two, trees conserve water and improve soil. Three, trees are windbreakers." His reverence for trees is infectious; one of his grandchildren is studying forestry in college.

Water, Paul says, "is my main challenge." Much of what he does on his farm is to lessen the impact of the changing climate. He is always on the lookout for crops that do better with less water, like drought-tolerant maize, which has proved to produce at least a minimum yield in the driest of conditions. Nearly two-thirds of Africa's maize is cultivated in drought-prone, water-deficient areas. Over the past decade, new drought-tolerant maize varieties have spread to cover more than seven million hectares in thirteen countries in sub-Saharan Africa. The scaling up has been particularly impressive in the Great Rift Valley; Uganda's maize production, boosted by drought-tolerant varieties, has increased to three tons per hectare from two tons, while Ethiopia has at times reached up to four tons of maize per hectare, rivaling the production of South Africa's modern agriculture sector.

"There are lessons to be learned here," says B. M. Prasanna, the director of the global maize program at the International Maize and Wheat Improvement Center, or CIMMYT, which backed Norman Borlaug's Green Revolution work. Africa's drought frequency, he says, is increasing. "There is now a serious drought somewhere in sub-Saharan Africa almost every year, particularly in the Horn and Rift Valley," he adds.

The Rift Valley has also been a major battleground in the fight against pests and crop diseases that have proliferated with the changing climate. Prasanna notes that climate extremes—drought, excess rain, unusually high temperatures—triggered six major crop epidemics in the past decade, causing billions of dollars in agricultural losses. In recent years, a swiftly moving plague of locusts devoured crops across the region, and maize lethal necrosis disease ruined many farmers' staple crop. Now farmers like Paul and Johnson are battling the fall armyworm, so named because of the way infestations march across a field, wiping out a crop in a couple of days. Moths fly at night, sometimes traveling one hundred kilometers in a single stretch, and lay up to one thousand eggs on young maize shoots and grasses. The larvae and worms that emerge spend their lives eating away. By 2022, seven years after their first appearance on the continent, the insidious forces of the fall armyworm had advanced to forty African countries, according to Prasanna's count.

The Ugandan farmers along the Nile note that the army-worms—which they call "Ami-worms" because they originated in North America—appear when the temperatures climb and only disappear when the rains come. So prolonged heat waves and sporadic rains often foil any efforts to control them, such as spraying with pesticides.

Which leads to another conundrum. "The pesticides may help immediate food security to secure a crop beset by pests," Prasanna notes, "but in the long run they can do more harm as pesticide use leads to biodiversity loss of beneficial insects and pollinators." Saving one crop thus leads to difficulties in growing other crops dependent on those pollinating insects to germinate and thrive. In West Africa, pesticides killed off many of the tiny chocolate midges that pollinate the cacao pods and support much of the world's chocolate supply. In the absence of the midges, many cacao farmers resorted to the far more expensive and time-consuming practice of pollinating by hand.

Paul and Johnson say they prefer to avoid using pesticides for these environmental reasons, but also because they can't afford to buy the spray and they lack the proper safety clothes and equipment to protect themselves from the chemicals. Their strategy, instead, is to diversify their farming away from traditional crops like maize and focus on growing more vegetables and hardier grains like amaranth, millet, and sorghum. They also sow other plants, like flowers and weeds, blessed with properties that act as natural pesticides.

Paul arrives at his half acre of amaranth, an ancient, highly nutritious grain prized for its antioxidant and medicinal qualities. He wades into his field, admiring the lovely deep-red grain heads.

"One, amaranth prevents malnutrition and treats diseases," he says. "Two, it matures quickly and makes quick money."

Paul and other farmers also see amaranth as an environmentally friendly alternative cash crop to sugar cane and rice. He is aiming to double his annual output to about one hundred kilograms, which he expects will bring in about $100 in sales. "I will keep expanding," he tells Dorothy. He hopes to buy a threshing machine; the current method of harvesting by hand is too slow and labor intensive.

Amaranth is in high demand along the Nile. It is a key ingredient in the porridge mix developed by Dorothy's nutrition program—amaranth and soybeans for protein, millet for iron, maize for energy. As the porridge has proliferated from one Nile village to the next, the widespread childhood malnutrition from two decades ago has largely vanished. Johnson, who grows amaranth and other porridge ingredients on his farm, has developed his own porridge mix to sell at the local market. He takes out his cell phone and pulls up his camera to show two photos. One is of a severely malnourished child with hollow cheeks. The second is of the same child after two months of being fed the porridge, cheeks now plump. "This," he says, "gives me the greatest satisfaction of all."

◈

Dorothy beams when she sees the noon lunch lines at the Namasagali primary school, just down the dirt road from Daniel Sebakakyi's farm. More than a thousand children gleefully run from their classrooms to the school's courtyard. Here, malnutrition is also on the run, in another of Dorothy's initiatives. With plates and bowls in hand, the children await a stew of maize, beans, tomatoes, eggs, and eggplant. It will be their second hot meal of the day. For breakfast, before the start of school, the menu was porridge made of amaranth, maize, millet, soybeans, sugar, and silver fish. The parents of the students supply the maize and beans and the fuel wood burning under big cauldrons. Everything else is grown here by the students and faculty; even the eggs are local, thanks to the chickens roaming the grounds.

This filling school-day diet is highly unusual for Uganda—and for most places in Africa—where students at government schools receive, at best, one hot meal a day, served in the morning. In 2018, before lunch was offered, 140 students, most of them boys, attended Namasagali primary. Four years later, with the commitment to serve two meals, enrollment has soared to 1,126 students, with girls in a slight majority, 566–560.

Parents are eager to send their children to school here; at home, they usually only prepare one meal a day. "In our villages, when families have lunch, they might not eat supper," says principal Uthman Aspa. "Now students get lunch here and then supper at home. They have more energy to study better. At other schools, you'll have hungry teachers teaching hungry students."

That was the case at Namasagali primary before Dorothy arrived in 2004. Back then, the school only tended a small garden, less than one-quarter of an acre. Now the twelve-acre campus includes six acres of crops and trees. Central to the curriculum is a focus on agriculture and its role in both nourishing the students and their community, and in preserving the environment. "This will be the central dilemma of our future. And our students will be prepared," says Uthman.

The materials for their outdoor classroom are the onions, collard greens, cabbage, eggplant, cassava, sweet potatoes, carrots, tomatoes, amaranth, millet, and other crops growing all around them in garden plots. The lessons unfold under the tall trees on the school grounds: mango, guava, acacia, eucalyptus, teak. At planting time, the students receive seedlings to take home to cultivate. They have installed a rainwater harvesting and drip irrigation system and constructed a greenhouse. They transformed the old outdoor volleyball and soccer field into an agroforestry laboratory, where fruits like avocados, jackfruit, and guavas flourish above vegetable cover crops blanketing the ground. Local farmers come to the school for learning field days. Officials from the state forestry service stop by as guest speakers, explaining how farmers can live with the forests. Beekeepers teach the ways to make honey.

"We don't grow sugar cane or rice here. We teach about agriculture in harmony with the forests and wetlands," says Augustus Baboola, the lead agriculture teacher. "We still have trees to learn from. The community was cutting down trees, so we planted them."

He leads the way on a tour of the fields, walking slowly in the wilting heat as the temperature soars past 90 degrees. We pass the chicken coop where eggs are laid, wander through the sack garden where cabbage, spinach, and onions grow in large burlap bags filled with soil (a demonstration on how to grow in small spaces), inspect the potato learning garden where experiments reveal which varieties do best in dry conditions, admire the stand of new *robusta* coffee trees, and arrive at a small teak forest, where the temperature in the shade immediately drops by several degrees. "You see the cooling benefit of the trees," Augustus says. "And the income possibilities. Teak is very good for furniture."

The gardens also provide good grounds for civics lessons and learning about the role of government and big business. "Wetlands need to be protected," says Uthman, "but it's still up to the government to deal with people earning a living in the wetlands." Students have noticed that some farmers are told to get out of the wetlands while others are given funding to grow in the wetlands. The teachers debate the case of a farmer who was imprisoned for having destroyed thirty acres of rice in the wetlands. But, they wonder, did he do the right thing? His intentions were to protect the environment.

There is also the story of a government concession to investors to plant bamboo trees in the wetlands. Members of the local community protested, suggesting that it would be better to replant eucalyptus trees, which grow in harmony with grazing livestock, provide shade and nutrients for fish ponds, and yield good wood for cooking fires. Before the bamboo thicket could get too dense to prevent grazing, the community burned down the trees. They grew angrier when they were told the bamboo was to be used in the manufacture of toothpicks for people in China and Japan.

"Toothpicks!" the cry arose. "How many people here eat with toothpicks? You only need toothpicks if you eat meat. Who can afford meat?"

Certainly not many of the families with children at Namasagali primary. Still, thanks to the morning porridge and the vegetables growing all around that end up in the lunch stew, these students have beaten the awful odds of childhood malnutrition in Uganda.

◈

When the Iowa State livelihoods program began in 2004, malnutrition was the underlying cause of more than 40 percent of all the deaths of children under the age of five in Uganda. Of those who survived, more than half were stunted in some manner by malnutrition, either physically, cognitively, or both. The clinical definition of stunting is low height for age. But in reality, stunting—particularly the stunting of cognitive development in early childhood—leads to a lifetime of challenges: fewer years in school and less learned when in a classroom, lower productivity and lower wages in the workforce, a greater chance of chronic illness in adulthood because of an immune system compromised by childhood malnutrition. In the communities along the Nile, some of the poorest in the country, the impact of malnutrition was even worse.

Many of the families of smallholder farmers experienced an annual hunger season that could last for weeks or months before the new harvest came in. Malnutrition rates soared during this time. Dorothy noticed how parents often took their malnourished children to the hospital, spending what little money they had in the desperate hope that doctors could save them. Her remedy would be for the farmers to provide the cure themselves with more nutritious foods.

One of those farmers, Jane Sabbi, speaks of those nightmarish days. "When you look at a malnourished child, who is waiting to die, the parents are waiting to bury them. . . ." Jane's voice trails off, lost in remembering. "I had seen so many severe cases. I lost count. . . ."

Then, suddenly, she snaps out of it with a burst of joy.

"And now they are alive!" she exclaims. "They've taken the porridge. The porridge saved them. We have saved many, many children."

When Dorothy first visited the village of Naluwoli in 2004, Jane was struggling herself as a smallholder farmer. She and her husband lived with their seven children in a mud house with a thatched roof that barely held back the rain. Malnourishment was everywhere in the community. Dorothy recognized Jane as an ideal candidate for Iowa State's agriculture improvement program, and Jane eagerly enlisted. She quickly applied the lessons to diversify her crops with new varieties, to improve her soil, to integrate trees into her farming. Soon, Jane, too, was a trainer. She first set off on foot to share all she was learning with her immediate neighbors. Then she got a bicycle and pedaled to more homes, expanding her route. Eventually, she received a motorcycle from the program to reach villages miles away. Jane, racing over dirt pathways in a red dress, became a common sight.

One day, she returned home with exciting news. Dorothy had asked her to take on another task, to become a health advisor in addition to being a crop trainer. After working with farmers to adopt new practices to preserve the environment, Dorothy wanted to make sure to address the problem of malnutrition. She dreamed of establishing a network of Nutrition Education Centers (NECs) throughout the Nile basin where pregnant women, young mothers, and babies could obtain a nutritious meal as well as nutrition education, porridge cooking classes, and health monitoring. The mothers, all farmers themselves, would not only learn how to grow food to heal the land but also to heal their children. After all, Dorothy reasoned, good health is the purpose of agriculture.

Jane could hardly contain her enthusiasm. "I have been told to go back to the homes and treat the children," she told her husband.

"Treat how?" he asked.

"We will treat with the food that we grow," Jane replied. "We will make it into a porridge and feed the mothers and children."

They would grow the porridge ingredients themselves. Amaranth, millet, maize, and soybeans in the fields, and milk from the cows. The grains and beans would be ground into flour and mixed with the milk. Sugar would be added for energy, as would the tiny silver fish from the Nile for additional protein.

"One mother came with a child, severely malnourished," Jane recalls. "Then another, and another. We started with three children that first day."

In the first week, 25 mothers came with their children. After the first month, there were 125. NECs proliferated through the Nile basin. Iowa State, working with Uganda's Makerere University, expanded the Community Nutrition Program to thirteen sites, with each NEC equipped with a gazebo meeting space and a kitchen and latrine. Often, the gatherings spilled out of the gazebos and into the shade of tall trees. The mothers and their children (or grandmothers who come with malnourished grandchildren in their care) enroll for at least six months while their health is monitored. When the upper-arm measurements indicate recovery from malnutrition, the moms receive training and seeds and other planting material to start their own gardens of nutritious foods, including the ingredients to make the porridge. In the dozen years since Dorothy and Jane opened the first NEC, more than 8,000 mothers and children were treated. In the summer of 2022, Dorothy calculated, about 1,000 mothers and children were visiting the NECs. Mothers showed off the recovery of their children and encouraged others to keep up the nutritional practices.

Where Jane led, many other farmer-mothers followed, nourishing children and preserving their land, doing their part to avoid the Great Collision. Winding along the Nile is an impressive trail of women farmers who joined in their twenties and thirties and are now passing on their knowledge and commitment to the younger generation.

Like Rose Mbiira, now nearing sixty, who remembers the depth of her misery. "Iowa State found me in the forest," she says. "We had no food, I had no income. We were eating once a day." Her husband had died and left her with three hundred acres of forest and wetland. He raised cattle, but passed them on to his older children from a previous marriage. Rose, who was sewing clothes in the local market, was terrified at the prospect of becoming a farmer, particularly of so much land. "I saw farming as a punishment, as a bad thing to do, such hard work. This land was all bush and swamp. What was I to do with that?" And then it got worse. "I woke up one morning and it was all flooding from the Nile. My house collapsed. The water stayed for three days, fish were all over. We thought the world had ended."

But her life as a farmer, and a conservationist, was just beginning. With the help of the Iowa State–Uganda livelihoods program, Rose planted crops in harmony with the trees and the water. From the bush has grown a wondrous garden, stretching over dozens of acres. Maize, cassava, tomatoes, sweet potatoes, peanuts, bananas, and the local crops sukuma wiki (a kale-like green vegetable) and amaranth. "I didn't know you could eat sukuma and amaranth," she says. Now everyone knows. Rose pioneered the growing of nitrogen-fixing soybeans in the area, which helped heal the soil and pack the porridge with protein. "I have found that it is all possible," she says.

Like Tapenensi Ngozi, who covered her three acres with the porridge ingredients and a thicket of fruit trees—mango, pawpaw, jackfruit, avocado—as well as grasses and other fodder to feed two dozen pigs. She joined neighbors and other women in the Nutrition Education Center to form a savings group, which helped to put her children through school. In 2021, she used some of the savings, and income from the sales of several pigs, to buy her first television. She was curious about the world and has since become an information center for her community.

"Now we know what's happening," says Tapenensi. One day in the summer of 2022, the news brought word of record heat in London. Tapenensi found it both disturbing and oddly comforting, realizing that people far away were also suffering from sudden climate extremes. "Oh, it's so hot there too," Tapenensi sympathized. "So it's not just here." Most days, there was news of conflict somewhere in the world. "We see the wars, the refugees, and wonder where Uganda is going," she says. She discovered the connection between the war in Ukraine and higher food prices in Uganda, learning from the news that Africa imports grain and fertilizer from Ukraine. "I don't expect prices to come down unless the war stops," she says. "When there is war, people don't grow food. It is dangerous to work in the fields." Her conclusion: Why do we depend on farmers from so far away? We should grow the food we need ourselves.

Tapenensi and her neighbors have also learned from nature documentaries, discovering the dangers of damaging the environment. "When you don't leave trees in the world, the rains stop. Even factories have emissions that reduce rain," she says. "For farmers, when you cultivate your land poorly, when you're not planting trees, not digging trenches to stop the water runoff, it leads to degradation, which makes farming harder. We must all know that we are doing this to ourselves."

Like Susan Namabiro, who despaired at the cutting of trees and set out to correct the destructive action of others. "We learned to plant trees in response to all the trees being cut for sugar cane," Susan says. She is sitting under a towering palm tree near her house, which is surrounded by four acres of crops and fruit trees. Two dozen chickens and ducks peck at maize kernels she has scattered about, and nine cows graze nearby. She was a young mother in her thirties, raising a house full of daughters, when Dorothy first came by. Since then, she and her neighbors have deployed new varieties of beans, grains, roots, and tubers developed with cutting-edge crop science by Uganda's agricultural research institutes, Iowa State, and international centers. And she has shared her knowledge with the community.

"When we introduced amaranth and mixed it with porridge, we took photos of the children before and after. The mothers could see the difference, they saw the importance of nutrition," says Susan. "They stopped talking about witchcraft making their children sick and understood it was malnutrition. And that they themselves could prevent it."

The same is true for the environmental changes bedeviling their farming. "The people growing crops in the wetlands are the ones bringing us the drought and unpredictable weather. They say they have no alternatives to growing there," Susan adds. "I tell them about the harm of planting rice and sugar cane, and that amaranth and millet are better for their children and their income. They do have a choice. We all do."

As the number of severe malnutrition cases have dwindled with the expansion of the NECs, they have transitioned to alumni support groups where women who once came with malnourished children now come with excited reports of their harvests and livestock, and stories of newly learned income-generating skills like sewing and basket weaving.

Mainly, though, says Jane Sabbi, "They come to say 'Thank you.'"

At a gathering under the huge mango tree in front of Jane's house, several dozen mothers and grandmothers show before-and-after photos of their children. They share stories of malnutrition nearly taking the lives of their children before the porridge helped them recover. They talk about how well their children are doing in school, how they are coping with the stunting side effects of childhood malnutrition.

"This is my favorite part, when the mothers come together as a support group and talk about their progress," says Dorothy. "What more can you ask in the way of sustainability?" At one support group gathering, mothers came with their cows, calves, and goats, purchased through their savings groups. A song arose as they shouted out the things they cherish: healthy children, recovered children, agriculture, trees, porridge, amaranth, millet, maize, soybeans.

Jane, in her fifties now, sits alone on the side of the gathering under her mango tree, listening. And dreaming of even greater possibilities. For, she says, she is just getting started. "I will be a businesswoman," she vows.

She offers a tour of her burgeoning farm, flourishing with an array of crops she learned to grow: bananas, cacao, and an assortment of vegetables, along with the amaranth, millet, soybeans, and maize for the porridge. There are chickens, pigs, and goats. She has a new house, a sturdy one made of bricks and concrete with a metal roof that doesn't leak in the rain. There are a couple of cinderblock sheds where she sorts and stores the porridge crops. She envisions adding a mill to grind the grain into flour—now she carries it down the road to a mill—and other machines to package it.

"I will start a business selling the porridge mix. I will put a label on it, listing the proteins, nutrients, everything that's in there," she says. She will put her brand name on the package.

"How about 'Jane's Flour,'" she suggests. She likes the sound of that.

And then she will market Jane's Flour in Uganda's big cities, and beyond its borders. "I will take it to Jinja, Kampala, even to Kenya," Jane says. "I will take it to schools. I want it to be available everywhere."

She and many other farmers in the Great Rift Valley are indeed just getting started. There are many more children, and one planet, to save.

CHAPTER 3

THE GREAT RIFT VALLEY: KENYA

Why Didn't We Do This Earlier?

ACROSS THE VAST DRYLAND savannas of Kenya, between the Ethiopian highlands and the White Nile of Uganda, drought is a constant threat to the farmers and their crops and livestock. Here, on a cool July morning, before the day's heat sets in, Kenyan farmer and pastor Pasqualine Mulusya seeks the bucolic land of milk and honey—or the milk at least.

"Slowly, slowly. Watch your step," Pasqualine cautions as she steps carefully across her field in south-central Kenya. "You can't see it, but we've terraced the land to hold the rain whenever it comes." She suggests a step-and-slide technique to traverse her sloping land. One step down and then slide on your heels, or backside if you must. We can't see the terrac-

Left: *Land of Milk and Honey*
Pasqualine Mulusya on her Kenya dairy farm

ing because the grass is so tall, almost waist-high. This isn't just any grass. It is a lush, thick variety of *Brachiaria*, favored by Pasqualine and her dairy cows, and by scientists striving to both increase the milk production and lower the methane emissions of those cows. Her meticulous journey through this meadow is one of the most important in our planetary quest to both nourish and preserve.

Pasqualine began her dairy operations about a two-hour drive east of Nairobi in 2018, with one cow and 160 grams of *Brachiaria* seed—40 grams of each of four varieties. She sowed the seed in several small patches and waited. Saplings emerged and then tiller shoots, which she nurtured across her land, first on a flat stretch and then down a newly terraced hill. Four years later, walking her farm, she enthusiastically shows off twenty cows and calves and the eight acres of grass that keep them fed all year long. And the cows, in turn, keep her family in milk for their tea every day of the year.

"We drink a lot of tea in our house, so we take a lot of milk," Pasqualine says. "We were spending 300 shillings a day on milk, 600 for a weekend." That's several U.S. dollars a day, a big expense in a country where the average household income is about $800 per month, and less in the rural areas. Pasqualine had been growing crops on her small farm, mainly the standard maize, which had been struggling due to the unreliable rain. She thought, why not get a cow and have a ready source of milk? So she did. "And then," she says, "Donald came around and gave us the seed."

Donald is Donald Njarui, a dairy fodder specialist at the Kenya Agricultural and Livestock Research Organization. He is Kenya's Johnny Appleseed of *Brachiaria*, handing out grass seeds wherever he goes. He is so popular, and important, that he was awarded a presidential medal—Elder of the Order of the Burning Spear—for outstanding service to the people of Kenya. "You know, Kenya has the highest milk consumption in Africa," Donald says. "If a family doesn't have milk in the house, they won't offer you tea. It would be a great shame for them to not have milk with the tea. Every Kenyan house has milk!"

And so, with every cup of milk-laden tea, the Great Collision intensifies. For the number of Kenyan tea-drinking, milk-consuming households is multiplying rapidly. The country's population, at about fifty million in 2023, is expected to more than triple by the end of this century. The population of the capital city, Nairobi, is expected to triple by midcentury. Which means the demand to increase milk production is also multiplying rapidly. Kenyans' per capita consumption of milk is about 110 liters a year (slightly above the global average) and has been steadily rising as the country's middle class expands, urban incomes increase, and awareness spreads about the importance of animal source proteins for combating malnutrition and raising healthy children. For these same reasons, milk consumption is also rising across Africa, although the continent's per capita rate of about 37 liters annually is the lowest in the world. Kenya's dairy sector is about 14 percent of the country's agricultural GDP. Nearly two million smallholder farmers, like Pasqualine, produce nearly 60 percent of Kenya's milk; the rest comes from larger commercial farmers. As for the cows, there are more than five million, producing about four billion liters of milk annually. Still, that satisfies only about half of the national demand; imports from neighboring countries fill the gap.

Kenya, and all of Africa, would like to be self-sufficient in milk. There are two ways to achieve that: increase the number of dairy cows, or increase the productivity of each cow. Increasing the number of cows has significant environmental consequences. Dairy cows, and other livestock, emit methane through their belching as they digest their fodder; they are among the largest emitters of methane, one of the main greenhouse gas contributors to warming temperatures, which in turn roast the crops in the field. This has dairy cows and other ruminant livestock often cast as villains in the discussion of climate change.

So, the great challenge looms: how to increase productivity of dairy cows while limiting their climate hoofprint? This is what made Donald, with his grasses, an Elder of the Order of the Burning Spear. *Brachiaria* outgrows most other grasses in Africa, with potential yields of thirty tons per hectare (about two and a half acres) per year. Farmers are able to harvest throughout the year, sometimes cutting the new growth as many as five times. It has proven to be adaptable to climate extremes, showing a tolerance to drought, flood, and shade. It grows on drylands and in swamps, and in poor, acidic soils as well. *Brachiaria*'s nutrition content can help double or triple milk production per cow, and scientists are intrigued by the possibility that it may include compounds that reduce methane emissions during digestion. It can grow for more than a decade without replanting, meaning the soils can store more carbon (each time a field is plowed, stored carbon is released into the atmosphere). Its deep root structure helps to improve soil health by minimizing nitrogen loss and lessening erosion. And it boosts farmers' income with increased milk production and the opportunity to sell surplus grass. Other grasses must be eaten fresh, shortly after cutting, or they lose nutritional value. Not *Brachiaria*; it can be consumed dry, meaning surpluses can be stored and stretched throughout the year, a particularly valuable quality during the ever-lengthening dry season.

Donald scattered the seed from farm to farm, wherever he found a receptive farmer. The project also set up a Village Knowledge Center in the town of Kangundo, where Mirriam Makato maintains a database on the *Brachiaria* forage-dairy value chain for farmers. The center has been a hive of activity during the drought. "The rain this year has been a disaster," Mirriam says. "Last season it failed, this season it failed."

"There's too much risk," adds Donald. "Farmers are looking for more stable seed varieties in the drought, so they can be assured they will at least have some kind of crop. They don't want high yield, they want stable seed, less risk."

In the center, farmers gather to swap seed experiences. Those without reliable access to the internet come to use computers to find information on agriculture and livestock farming, and access smartphones to connect with a WhatsApp community among farmers, extension agents, scientists, and meteorologists. Mirriam leads group discussions for farmers who are illiterate. After Russia attacked Ukraine in 2022, destabilizing global grain and fertilizer supplies and prices, farmers poured into the center in search of the best prices and markets.

"In one month, we got 250 messages from farmers on WhatsApp," Mirriam says. It was a WhatsApp campaign supported by *Brachiaria* farmers in seven counties that nominated Donald for the Order of the Burning Spear award. "Not even my director has this award!" he says with a burst of laughter. But he credits his inspiration for the center to a meeting he had in India with M. S. Swaminathan, who worked with Norman Borlaug during the Green Revolution. "He talked about empowering farmers with knowledge," Donald says.

The knowledge center, and the *Brachiaria* campaign, also follows the lead of the late environmentalist Wangari Maathai,

Kenya's own Nobel Peace Prize laureate. Her teachings still stir the hearts of many Africans: "You cannot protect the environment unless you empower people, you inform them, and you help them understand that these resources are their own, that they must protect them."

Pasqualine understood this immediately upon meeting Donald and quickly seized the opportunity to be a *Brachiaria* pioneer. As a pastor at a local church, studying for a PhD in theology, she was eager to become a better steward of God's creation. She would save souls in the church and save the planet on the farm. She also wanted to do better on her other biblical goal: feeding her neighbors. While living in the nearby town of Kangundo, she was mainly a "telephone farmer," calling in instructions to workers. In 2014, she moved into the farmhouse to become a more active farmer in between the sermons.

At the time, she had one dairy cow, and a patch of Napier grass to feed it. She was also growing maize and beans, the standard crops, but the increasingly frequent drought conditions decreased yields. She also had some banana plants, but they didn't fit with her dairy ambitions. "Cows don't eat bananas," she says. County agricultural extension agents encouraged farmers to switch to greenhouse farming to neutralize the climate impacts. Pasqualine tried strawberries. She installed an irrigation system and hired workers to help with the planting and picking. Then the local marketer abandoned the project. "All was lost," she recalls. "You know, with strawberries, when you can't sell them, they spoil. After one season, that was it."

She feared the farm was failing, and she and her husband Joseph worried about its future. Their six adult children were living in the city with little interest in becoming farmers themselves.

Then Donald came by with the *Brachiaria* seeds. "It was a

godsend for us, and for our farming livelihood," Pasqualine says. "This grass saved our farm."

The rains have been sparse for the past three seasons, the drylands becoming ever drier. But this preacher rejoices as she wades through her acres of hardy grass. "We are able to harvest throughout the year. We start at the top of the hill and work our way down, and then go back up to the top again." *Brachiaria* can grow past five feet tall; wielding a machete, Pasqualine cuts it down to several inches above ground, which stimulates regrowth. *Brachiaria* is ideal for a cut-and-carry system; the farmers cut the grass and haul it to the cows, rather than having the cows graze and munch the grass down to the soil, which makes regrowth more difficult. It also reduces the pressure to expand into forests for more grazing land, and limits land degradation. Pasqualine harvests about thirty tons of grass every month, which leaves some surplus to sell on the market. In the pre-*Brachiaria* days, she often needed to buy hay to supplement her more meager on-farm harvests.

The increased production and lowered expenses from the ready supply of grass has allowed her to expand her herd and modernize her operations. The chatter from a grass-cutting machine breaks the pastoral stillness. "The machine cutting makes the grass easier to eat. It's less work for the cows which means more energy to produce milk," Pasqualine says, raising her voice to be heard above the din. Nearby, workers mix the grass with other fodder to provide additional nutrients. The milking is done by machine, three times a day. On a blackboard, workers record the amount of grass consumed and the milk produced by each cow—Mary, Lena, Dora, Zawadi, Moreen, Smart, Ramos, Lily, Ricarda, Brown, Victoria, and Blessing. The chalk markings show that with the *Brachiaria*, Pasqualine's cows have doubled milk production, to as much as twenty to twenty-five liters each on some days.

Still, that's only a glass half-full compared to milk production in the U.S. and elsewhere. Pasqualine's next goal is to improve her herd with higher-producing breeds. "You could get up to forty liters per cow," Donald says. More liters per cow means less methane per liter. Pasqualine smiles at that equation. Nourishing and preserving.

"God willing," says the minister-farmer.

◈

That is also the goal, and prayer, of Janet Kisyanga. Her per-cow milk output has similarly doubled, to upwards of twenty liters a day since Donald arrived at her farm with the *Brachiaria* in 2018. And her cows and calves have multiplied from two to eleven. She sells her milk at a shop on the edge of her farm, and she can't keep up with the demand.

"The milk from the cows feeding on this grass is very thick, very heavy, the flavor is very sweet," Janet tells Donald. "The customers tell us the taste is very different. They love to put it in their tea. People can tell the difference."

Janet is up before the sun every day, beginning the first milking at four in the morning. There will be another near midday and then again at six in the evening, as the equatorial sun begins to set. Janet, too, records the grass consumption and milk production of each cow; Baraka, which means "blessing," usually sets the pace.

Janet's son, Daniel, home from college, fills the feeding trough with grass and sprinkles in some maize flour. "It calms them," he says of the flour. With the country's increasing demand for milk, he envisions his future as a dairy mogul. He is studying animal health and nutrition at the vocational school about twenty-five miles from their home. The greatest lessons, though, come outside the classroom as he tends the cows with his mother and his father, Samuel. With each shovelful of grass, he calculates the value of the *Brachiaria* for animal and environmental health.

"If we can get thirty liters per cow per day, that's a good thing," he says. "And it's quick money." While his classmates aim for jobs in the city, Daniel says, "I want to be a farmer."

"How many cows do you want?" Donald asks.

"Two hundred," Daniel says.

"Oh, you will take over the farm then."

"Maybe," says the son, looking hopefully at his mother.

"It's hard work," she says. Particularly as the drought stretches on.

Janet explains that the income from the increased milk production has enabled the family to construct an elaborate water harvesting system on their homestead that benefits all who live here: the family, the cows, the grass, the vegetables. When the rain comes, not a drop goes to waste.

Eaves on the metal roof of the house and cow shed funnel the rain into barrels, which are connected by pipes to a larger holding tank. This is used for drinking water, after it is treated with chlorine. Much of the ground around the structures is paved, as are the pathways leading to the fields. The rain that falls here is channeled into thin canals that run downhill into a water pan, which has expanded over time to the size of a small pond. From there, another set of pipes carries the water down a slope to a field of crops. A network of plastic drip irrigation tubes laid in the soil brings water to the tidy rows of vegetables: red, green, and yellow bell peppers, tomatoes, melons, pumpkins, bananas, Swiss chard, varieties of local greens, and amaranth. The paths surrounding the field are lined with cow manure, laid out to dry before fertilizing the soil; not an ounce of that precious dung goes to waste, either.

Beyond the vegetables grow a couple of acres of *Brachiaria*. Janet has integrated trees with the grass; the acacias with their flat-topped awnings provide shade that reduces evaporation of the morning dew, while the roots aerate the soil and provide needed nutrients. "You can see the grass does better growing closer to the trees," Janet says. "It's greener, taller."

As she walks past the pond on a hot July afternoon, she laments that the water level has dwindled to the one-quarter mark. "We haven't experienced a period like this before, low rains for three seasons in a row," she says.

Janet asks Donald when the next rains will come.

Donald replies with a question. "When did it last rain here?"

Janet laughs. "I think it was March."

"No rain in May?"

"Oh, May maybe. No, April," Janet decides. "Yes, it rained in April." Nothing in the three months since.

Donald says the next rain should arrive in October, at least according to traditional weather patterns. "Those rains are more reliable," he says. Except for the past two years, when they weren't.

Janet sighs. "We have hope," she offers. "God provides." Alas, there was no substantial rain in October, nor in the subsequent months. It was not until the following spring when the drought broke, temporarily.

Janet follows her *Brachiaria* field down to a narrow dirt road that runs past her farm. There stands King's Health Center. On the wall is a message from the Book of Exodus: "I am the Lord Your Healer." It was God's promise to the Israelites as they made their way to the land of milk and honey.

◈

Donald resumes his *Brachiaria* grass evangelizing. He climbs through an opening in a barbed wire fence and hails a farmer in the distance. Martin Kisuke hurries over, moving swiftly for his seventy years. He smothers Donald in an embrace. "Thank you so much for all you have done," he says. "You'll see, I'm expanding the grass, replacing the maize. I'm through with the maize."

Martin reports his farm had only four days of rain in the past two months. "I'm not a meteorologist, but something is happening," he says. His patch of maize looks miserable, the green husks having turned yellow and then brown for lack of rain. The stalks slump from heat stroke and thirst. The cobs are small and malformed. "Normally, we should be harvesting in a couple of weeks," he says. "Not this year." Next to his maize, a few rows of pigeon peas, usually more reliable in a drought, look equally pitiful. "Nothing," he says, summing up his expectations for the harvest.

But his *Brachiaria* grass, nearly two acres of it, is standing tall and green. Three workers are bending deeply, swinging their machetes as they harvest a long, narrow strip of grass. There is grass to cut most days of the year.

"Where I'm standing, this used to be maize," he reminds Donald. "The first year when you came, we planted grass to that post." He points to a wooden electricity pole about fifty feet away. "Then, we planted to that mango tree." He points another fifty feet farther in the distance. Next year, without the maize and peas, it will all be grass, an additional acre or two. He figures he won't have to worry so much about the weather. He is more relaxed and has scaled back his extravagant appeals to God.

With the maize, he says, "I would offer a goat for sacrifice under the big tree on the hill, asking for good rain if God is angry with us." Now, feeling more secure with the *Brachiaria* as an ally against the changing climate, he says, "we just go to church."

As a staple of the Kenyan diet, maize is grown on almost every farm. Going forward, Martin says, he will buy his maize in the market. He runs through the numbers: for the past couple of years, with stingy rains, his plantings yielded a measly 100 kilograms of maize, with a value of about 1,500 Kenyan shillings (about U.S. $12) minus his costs for seeds, fertilizer, and labor. As he expands his grass cultivation, Martin believes he can harvest up to 900 bales a year, with each bale selling for 300 shillings. It's simple math; he comes out way ahead with *Brachiaria*. "I will now grow grass and buy maize to feed my family," he proclaims. "And I don't have to worry about rain for my maize."

Martin has few cows himself, so he sells the majority of his grass. With the income, he bought a baling machine, which he learned about through the Village Knowledge Center's WhatsApp group. The baler lightens the workload, packing the grass into neat bales of about fifty pounds each. Each bale, he says, contributes to the education of his family. He has seven adult children—"I've got a high school principal, a teacher, an accountant"—eighteen grandchildren, and three great-grandchildren. He says one grandchild will soon be graduating from college, setting an example for the others. "They like being on the farm," he says. "When I retire, I hope they will want to take over."

He has educated them beyond school as well, explaining the value of the *Brachiaria*—not only for the income, which helps assure them they will have a farm to inherit, but also for the environment, the protection of the land, the soil, the water. He follows Donald's evangelizing example, sharing some of his grass saplings with his neighboring farmers. He wants them to share in his gain and also thrive in the face of drought and heat. "Sometimes," he reminds them, "God needs our help too."

Martin hugs Donald again as they say goodbye. And again he says, "Thank you for all this. Thank you for bringing us the grass."

The old farmer also is thinking of his legacy, grateful that the grass has given him one to pass down through his family. If he has any regret, he says, it is simply this: Why didn't we do this earlier?

◈

Sita Ghimire, who was the principal scientist of the feed and forage development program at the International Livestock Research Institute (ILRI) in Nairobi and often traveled the countryside with Donald, knelt in Martin's field of *Brachiaria* and ran his fingers through a few strands. "This," he said, "is the variety that was taken from Uganda many decades ago and sent to Australia."

How it ended up back here, flourishing in the Rift Valley fields of these dairy farmers, is a curious, circuitous tale that winds back to Africa's colonial era, when natural resources like gold, diamonds, and timber were routinely extracted from the continent but rarely developed and sustainably enhanced at the source. It answers Martin's question: Why haven't we had this grass before now?

Brachiaria, Sita explains, is indigenous to Africa. It is one of the few forages suitable to the climate around the equator and throughout the Great Rift Valley, and is capable of sustaining livestock during the dry seasons. Throughout history, though, it was never domesticated, as African farms favored Napier grass, also known as elephant grass. At one point, *Brachiaria* was used for bedding on slave ships, which is how it made its way to America. Evaluations of the grass for use in pasture improvement in some parts of Africa began in the 1950s, but it wasn't cultivated and developed commercially on the continent. Researchers in Australia heard of the studies and saw the grass on a visit to Uganda. They took samples of the gene material back to Australia and introduced it to the outback drylands there. From Australia, *Brachiaria* varieties made their way to the Cerrado of Brazil and elsewhere in South America. It wasn't until 2014 when this African grass that flourished around the world finally came back home. Some improved cultivars from South America were shipped back across the Atlantic Ocean, this time as precious cargo rather than stuffed into bedding, and planted in Kenya and Rwanda. One of the first patches to be cultivated was still growing strong at the entrance to ILRI's Nairobi headquarters.

Since then, Donald and Sita, as leading purveyors of *Brachiaria*, have been in top demand in various parts of the continent. Their motto: Have seeds, will travel.

"Feeding livestock is a big deal in Africa," Sita says with grand understatement.

That is because livestock of all varieties are themselves a big deal in Africa. Of the estimated 1.5 billion people on the planet whose livelihoods depend to some extent on livestock—be it cattle, sheep, goats, pigs, poultry, or camels—more than half are in Africa. Globally, livestock generates about 40 percent of agriculture production value; in some African countries, it can be as high as 80 percent. ILRI estimates that Africa rears about one-fifth of the world's cattle; the continent's herd predominantly consists of indigenous breeds and is largely in the hands of smallholder farmers and pastoralists. In more than half of those households, women tend to the livestock, empowering them within their families and communities.

For these households, their livestock, particularly their cattle, are far more important beyond a source of income and nutrition. It is a close relationship that has stretched across centuries, first illustrated in ancient rock paintings and carvings. As ILRI describes in its glossy publication, *The Story of Cattle in Africa,* cattle are often the most valuable asset of a family, deployed in a variety of ways that are central to everyday lives and cultures. They provide labor and power for plowing the land to grow crops and then transporting the harvested crops and other goods to market. Their manure is an important nitrogen-rich soil fertilizer as well as useful material for fuel and construction. The hides and horns provide clothing, tools, decorative items, and ceremonial objects.

Perhaps most important, the ILRI story continues, "In Africa, cattle are an indispensable means for achieving savings and provide a traditional social safety net that helps to spread and manage risk and, in so doing, building household resilience to economic shocks. Cattle are often central to social exchanges that create social capital and redistribute wealth." In weddings, cattle are often the central prize in the dowry paid to the wife's family. They are given in celebration of births and in condolence of deaths. Cattle serve as both currency and a measure of social prestige. Also, ILRI notes, "The animals represent a form of long-term insurance, providing an asset to be sold, traded or borrowed against in times of necessity. Cattle are exchanged to bind families or to compensate for crimes and losses."

Putting their value in modern financial terms, ILRI director general Appolinaire Djikeng says, "Livestock are the ATMs of smallholder farmers."

In many African households, the livestock are part of the family. During one visit with Tesfaye and Hagirso in Ethiopia, they invited me inside their round *tukul.* With no windows, it was pitch-black inside; it took my eyes a few minutes to adjust. I felt something rubbing against the back of my legs. When I turned and looked down, I discovered it was their one cow, lying on the dirt floor with her calf. Tesfaye explained that it was a particularly hot day, so he invited them to come inside as well to rest and cool down.

As the saying goes at ILRI, "Where you find people, you find cattle."

You also find them at the center of contentious climate change debates. Campaigns to save the planet by reducing methane emissions call for an elimination (or severe reduction) of meat from our diets and major reductions in the number of livestock around the world. ILRI argues that in many parts of the world, livestock provides desperately needed nutrition, as well as income and livelihood resilience, in the face of environmental stress and climate extremes, particularly rising temperatures and drought. Africa's indigenous breeds, ILRI notes, are "well adapted to a remarkably wide range of environments—from the harsh fringes of the Sahara Desert in North Africa, to the drier areas of the Horn of Africa, to the wet tropical lowlands found along the Congo River, and on to the vast savannas of Southern Africa." They have become efficient producers of meat and milk on marginal land and in extreme weather conditions. Over the centuries, ILRI scientists point out, African livestock have developed thick skin to resist insect bites, short hairs to discourage parasites, a greater tolerance to diseases, and strong legs and hooves that enable them to walk long distances to scarce water and forage sources. They also mitigate food waste, a major contributor to greenhouse gas emissions, by consuming the bulk of plant growth that humans don't eat. While people eat only the ear of corn, for example, livestock feast on the rest of the stalk as well.

Still, in a grand twist of the Gordian knot, livestock is clearly a major source of greenhouse gases—through methane emissions from digestion of fodder and uncontrolled grazing that leads to deforestation and diminished carbon sequestration—and is thus a leading cause of the very same rising temperatures and extreme climate events that endanger their own health and the health and livelihoods of the farmers. Heat stress was a prime factor in the death of more than seven million livestock across the Great Rift Valley and the Horn of Africa during the epic drought at the start of the 2020s—the very same one that had Pasqualine, Janet, and Martin looking heavenward for rain. The UN Office for the Coordination of Humanitarian Affairs also reported that the resulting drop in milk production from the deaths of so many dairy cows left more than seven million children in the region severely malnourished.

According to the Global Methane Hub based in Santiago, Chile, 42 percent of methane emissions come from the agricultural sector, with 36 percent coming from fossil fuel extraction and 18 percent from waste (a large portion of that waste being spoiled or discarded food and its packaging material). Of the agriculture emissions, largely from livestock and rice, 80 percent comes from low- and middle-income countries, and the majority of that comes from smallholder operations. ILRI estimates that livestock accounts for more than 70 percent of Africa's greenhouse gas emissions. The Hub, which monitors and supports innovations and investments in reducing emissions, maintains that reducing methane gas is the fastest way to address climate change in the short term. In 2021, the United States, European Union, and over 110 countries signed the Global Methane Pledge to reduce global methane emission and keep the goal of limiting warming to 1.5 degrees Celsius within reach.

Yet, explains Hayden Montgomery, the Methane Hub's agriculture program director, "It's not just about curtailing livestock and agriculture productivity. We still need to feed the world. It's about improved feed and forage and animal husbandry." And, he argues, it's mainly about engaging smallholder farmers. "If we ignore smallholders, we won't improve mitigation." It's also about recognizing a global commons cost. "Animals emitting methane is a cost to all of us," he says. "Can there be financing spurred by a commons cost?"

But in the early rush of global financing for mitigating the forces driving climate change, the smallholders, as usual, are being left with the relative crumbs. Much of the money is chasing breakthrough technology—concocting seaweed and other exotic nutrient additives for forage, for example, or developing a vaccine for livestock that will reduce methane creation during digestion—that will likely be years in the making. For smallholders like Pasqualine, Janet, and Martin, these breakthroughs will be too expensive, and arrive too late.

"The best time to start was last decade," says Hayden, who grew up surrounded by dairy cows in New Zealand. "The next best time is now."

The rising clamor has increased the urgency of ILRI's work to transform smallholder livestock systems for the good of the environment while preserving the billion-plus livelihoods and social threads that depend on it. This is livestock's own Great Collision: meeting an increasing demand for meat and dairy production—particularly for the many communities in low-income countries that currently under-consume animal-sourced proteins and suffer from widespread malnutrition and childhood stunting—while using less land and with fewer numbers of livestock.

One way to accomplish this is through breeding, mixing the continent's current stock with more productive cows from Europe, North America, Australia, and New Zealand.

This demands a precarious balancing act between breeding to improve production while retaining the many positive traits of Africa's indigenous breeds that allow them to survive in the continent's harsh environments. To safeguard this genetic diversity, ILRI curates the Azizi (meaning "precious treasure" in Swahili) Biorepository at its headquarters, where five hundred thousand samples of livestock genetic material is kept in an ultra-low cold storage system.

A second way to address the dilemma of livestock is to reduce methane emissions by improving the fodder the cows eat, like *Brachiaria*, and thereby reduce the environmental hoofprint of livestock. And so, beyond the Azizi lab, in the Environment Center, ILRI scientists are measuring what exactly those emissions are, seeking to determine whether the impact is less from cattle raised in the open rangelands by smallholder farmers in Africa than from cattle raised in more confined settings at huge commercial operations in the United States and other richer parts of the world. Here, cattle and sheep shuttle between outside open-air pens and twenty-four-hour stays in glass-enclosed indoor chambers where the methane emissions from their bodily functions are measured. One set of tubes brings in fresh air while another pulls out the emissions to be analyzed by a series of machines and computers. Emissions from the manure and urine are separately measured, as is the impact of variable climate conditions and the different types of forage the livestock eat. In various testing climates with diverse African livestock breeds, early results indicate that the emission levels of methane and nitrous oxide in rangeland and smallholder farmer settings are lower than the estimates of the International Panel on Climate Change applied to global calculations.

More emissions studies are being conducted at ILRI's large

Kapiti Ranch out in the Machakos drylands not far from the dairy farmers growing *Brachiaria*. Kapiti, which means "place of hyenas" in the local language, is a wildlife conservancy designed to demonstrate that livestock like cattle, sheep, goats, and camels can live together in the same ecosystem with wildlife like giraffe, zebra, hyenas, wildebeest, gazelle, antelope, and leopards (as well as one lion), for the benefit of the environment and the land. If it works here, Kapiti could eventually become part of a vast contiguous wildlife corridor running from the Nairobi National Park south to the Amboseli wildlife preserve on the Tanzania border in the shadow of Mt. Kilimanjaro. Success here would create a valuable model for many African countries, particularly those in the Great Rift Valley, where the pressures of human population growth, agriculture expansion, and wildlife habitat conservation are pushing all living creatures closer together than ever before.

Ilona Gluecks, Kapiti's peripatetic manager, clambers to the top of a rocky outcrop to observe the herds moving across this breathtaking expanse. As dusk approaches, she explains how the research can help smallholder livestock households become better stewards of the environment. During the day, the livestock roam freely, and at dusk, herders bring them into fenced-in bomas to protect them from nighttime predators. Spindly metal towers, monitoring air quality, rise above some of the bomas like the antennae of giant insects. Sophisticated gadgets affixed to the antennae measure the levels of methane and carbon dioxide gases over time, seeking answers to two critical questions: How long do these greenhouse gases linger in the air after the bomas have been broken down and moved to new locations? And how quickly does the methane disperse as the manure dissipates into the soil? Ilona says early data indicates that if farmers move the cows to different pens every ten days in the dry season, it takes about ten days for the methane traces to wind down. Otherwise, if the cattle constantly gather in one place, the manure builds up and it may take years for the methane emissions to disperse.

Other instruments placed around the ranch measure carbon and moisture levels and vegetation growth. Ilona says the animals, in their grazing, become helpful agents in restoring degraded land. Their hooves break up soil that has become hard while baking in the heat; as the cattle and wildlife move around, they naturally mix the nitrogen and other nutrients in their manure into the soil. "It isn't long before we see the grasses came back," Ilona says triumphantly.

She drives along the edge of Kapiti where a fence separates the wildlife and livestock from a large commercial farm growing a variety of crops. She has also started several demonstration plots of her own, intercropping maize, beans, and other vegetables with grasses. She plans to host a field day where a couple hundred neighboring farmers would be invited to see how such an integrated agriculture system can help with the rehabilitation of degraded rangelands and also provide nutrition for both families and livestock.

Finishing her daily rounds of the ranch, Ilona settles down at a watering hole where cattle drink alongside zebra, gazelle, and wildebeest. In the distance, several giraffe approach. Kapiti's giraffe population stands at about 180 and counting; this is one of the few places in the wild, Ilona notes, where giraffe numbers are rising.

"This is the closest you can get to a pastoral system, with wildlife and cattle together," she says. "If your grandchildren want to see giraffes in the wild, not just in a zoo, we have to preserve this."

She reaches for her binoculars and studies birds arriving at the watering hole. There are pelicans, and a majestic African crowned eagle. "Haven't seen those here before," she says. Her vision of Kapiti's future, of livestock and wildlife and humans and the environment thriving together, benefiting all, continues to come to life.

"If we can achieve this, I think we've made a contribution. It must be possible," she says. "It must be possible."

◈

Leaving the ranch, on the two-hour drive back to Nairobi, the need to achieve Kapiti's potential becomes even clearer. Signs of the relentless march of urbanization are everywhere. New housing settlements on the drylands are popping up like desert flowers after a rain. The outlines of a new high-tech hub—a Silicon Valley of the savanna—are taking shape. There is speculation of a coming airstrip to handle the cargo traffic overflowing Nairobi international airport. Traffic jams clog the highway. The tripling of Nairobi's size, from its current five million inhabitants, is underway, gobbling up the farmland that has been nourishing the current population of the city.

It has many wondering: where will the food that feeds the city come from in the future?

"Let me show you," says Mike Munameza. He leads the way through the narrow passageways of Viwandani, one of Nairobi's densely populated informal settlements, a sprawling maze of dwellings hastily and crudely constructed from of all sorts of material—corrugated iron, wood, brick, cardboard. Smoke rises from cooking fires in many of the doorways. Garbage lines the passageways, streams of sewage trickle underfoot. Viwandani, bordering Nairobi's industrial zone just beyond the city center, grows by the day, by the hour, as people pour in from the rural areas seeking jobs, income, and a better life. By 2030, it is expected that more than two-thirds of Nairobi's residents will live in urban slums like this, a population proportion that is repeated in the largest cities throughout sub-Saharan Africa. This rural–urban migration is bringing the hunger of the countryside into the cities; about 50 percent of children living in Africa's poor urban settings are malnourished as their parents are unable to afford the food in the markets, says Elizabeth Kimani-Murage, a public health nutrition specialist and research scientist at the African Population and Health Research Center in Nairobi. Her studies have found that about 80 percent of households in Kenya's capital city are food insecure, scrambling daily to put enough food on the table. Here, the Great Collision between nourishing and preserving is visible around every crowded corner.

In Viwandani, Mike nimbly steps around the garbage and the sewage. "This area was a big waste site," he explains, describing a community dump that became a repository for all manner of human waste. "There's not enough toilets in our community," he says matter-of-factly. The residents have named this neighborhood Sinai, after the mountain at which the Ten Commandments were given to Moses by God, a place of revelation.

Which is what this Sinai becomes as Mike approaches a barrier of corrugated metal sheeting standing at the end of one of the alleys. He leans against a jagged, rusty door, pushing it open to reveal a wholly unexpected world of wonder. Mike steps across the threshold like Dorothy arriving in the Land of Oz. Drab and desperate turns to technicolor possibility. The garbage and sewage are gone. The paths are clear. Green plants grow everywhere, flowers bloom. Passion fruit vines dangle overhead. It is another veritable Garden of Eden—like Abebe Moliso's verdant farm in Ethiopia and Paul Babi's four-story tree canopy in Uganda—where you would least expect it.

"Clean and Green is our Dream," announces a splash of

graffiti on the wall. It is signed by "Sinai Ghetto Shinners" (pronounced "shiners"). That is Mike's organization of young community activists. "We're the shiners of the community, the cleaners," he says. The group formed in 2018 as a soccer club and community improvement organization. In 2021, during the Covid pandemic lockdown, they became urban farmers. "We decided to pick up the garbage and turn it into an urban farm, to plant some crops, vegetables, and raise poultry and rabbits," Mike says.

Just inside the door, an urban farm flourishes, running along the cinderblock wall of a timber factory. Kale, spinach, and an array of indigenous leafy greens, including sukuma wiki (a Swahili phrase which means "push the week," as it is seen as food to get you through the week) form a green wave; amaranth grain grows here too, and cowpeas, as well as bananas, tomatoes, pumpkins, and gourds. Given the narrow spaces, much of the farming is vertical, with one layer of plants spiraling above another. Their roots expand in raised soil beds made from an ingenious variety of materials. Some are constructed with scavenged wooden boards; others are fashioned from plastic water jugs or old blue jeans. In the "jeans garden," the bottoms of the pants themselves are planted in the dirt; the legs are filled with soil up to the waist, so the jeans stand like pillars. The vegetables rise out of the top, above the waist; vines poke out of holes in the pockets.

"We want to show people you can plant with anything," Mike says.

At the end of the garden alley, Mike pushes open a blue metal door, and enters the livestock farm, an alcove of chicken coops and rabbit hutches. There are twenty-five chickens, which supply the community with eggs, and a dozen rabbits, which provide a good source of protein. "Rabbit tastes like chicken," says Mike. "Actually, it tastes sweeter, fattier." Buckets under the coops and hutches collect urine and poop, which becomes fertilizer for the garden plants.

As the vegetables and animals flourish, so does free expression in the colorful graffiti and murals painted on the walls. The artwork praises the drive for food sovereignty and embraces *Ubuntu*, the African spirit of equality and collective humanity. It rejects all those who would use food for oppression. Old colonial masters and rulers of current regimes come under direct fire from the artists.

The Right to Feed Oneself is Dignity, is a ubiquitous slogan.

On the cinderblock wall behind the passion fruit vines, artists have drawn a tree mural with aspirational words hanging like fruit from the sweeping branches: *Unity. Dream. Love. Joy. Nature. Explore. Hope. Peace. Respect. Food.*

Messages surround the base of the mural, like life-giving roots: *Thy Tree Belongs to Community. The Greatest Fine Art of the Future will be Making a Comfortable Living from a Piece of Land. What You See Depends on How You View the World: To most People this is Just Dirt, to a Farmer it's Potential. Zero Hunger, Zero Pressure to Find Food. Ubuntu=Sharing. All the World's Problems Can be Solved by a Garden. Gardening is the Work of a Lifetime, It Never Finishes. Green is Life, Youth, Growth. Growing Your Own Food is Like Printing Your Money. The Soil is the Gift of God.*

On a corrugated metal wall, beside the chickens and rabbits, the graffiti is more defiant and righteous: *Never Lick Their Soles to Win Their Souls. Walk Like a King or Like You Don't Give a Fuck Who the King Is. Under Pressure But We Never Blast.*

On the far end of the alleyway, before the farm yields again to the desperation of the settlement, a Kenyan flag hangs above a broken tennis racket. The flag is a tricolor of black, red, and green horizontal stripes with two crossed spears and a Maasai shield at the center. "The green of the flag," says Greg Kimani, "that's this garden right here."

Greg is one of the leaders of Nairobi's urban agriculture and food sovereignty movement and the founder of City Shamba (which means "farm" in Swahili). "If you don't have money, you don't have food, and that is the thing that we are trying to demystify," he says. "We want to show that you can grow your own. That is our mantra." He literally wears it on his sleeve, and on his chest. "Grow Your Own" is emblazoned on his t-shirt.

City Shamba and the Sinai Ghetto Shinners are two of the grassroots organizations supported by the Cool Waters Nairobi initiative launched by the African Population and Health Research Center (APHRC). In 2021, it was named one of the ten finalists for the Rockefeller Foundation's 2050 Food System Vision Prize, which encourages innovative work toward creating regenerative and nourishing food futures.

"The Maasai people were the original inhabitants of Nairobi. There were streams and rivers and the water was clean and cool. And that's why they called it Nairobi. It means 'a place of cool waters,'" says Elizabeth Kimani-Murage of the APHRC. "It has green spaces, people are healthy. But of late, this is changing."

For the past two decades, she has been studying, with increasing alarm, the impact of urbanization on Nairobi's health, environment, culture, and spirit. It is far from the place that inspired the Maasai; with its crowded settlements, shrinking green spaces, polluted rivers, steaming garbage dumps, and ever-more people jostling for jobs and food, today's Nairobi barely resembles a place of cool waters. "When I was growing up, sharing food was a common thing that we did. It is about the value of *Ubuntu*. I am because we are. It is the spirit of helping one another," Elizabeth remembers. "But the world is urbanizing and we are losing that culture."

As she watches the concrete of the city spread out over surrounding agriculture fields, she stirs impatiently with an urgent imperative: the city must feed itself. "We need to start thinking about promoting agro-ecological urban farming so people can grow safe food and feed themselves. To restore Nairobi to a place of cool waters."

This mission grew in urgency and purpose as food prices in the local markets skyrocketed due to the supply chain interruptions of the Covid pandemic and shortages resulting from the war in Ukraine. At the same time, the food production from rural counties surrounding Nairobi continued to dwindle as the drought tightened its grip and the concrete jungle claimed more land. Suddenly, prices for vegetables, fruits, maize flour, cooking oil, eggs, and milk doubled and tripled. Hunger and malnutrition escalated in Nairobi, in Kenya, in Africa, in all the world.

"It is a challenge that many people are facing," Greg says as he makes his way past a smoldering, foul-smelling dump site where both people and cows are scrounging for discarded food bits. It is a scene that defines their Right to Food project, which frames their goal as "recognizing that the right to food is not a right to be fed, but the right to feed oneself in dignity."

A poster plastered on a wall across from the dump advertises an upcoming sermon on "How to be Rich in Africa, and other Secrets of Survival." As he walks past, Greg continues: "If my neighbor cannot have food, none of us can be food secure."

Just beyond the garbage dump, Greg greets another group of urban farmers. Along with a jeans garden, they have conjured two additional farming innovations. They have planted a "TV garden," where carrots, tomatoes, beans, peas, and spring onions sprout from discarded television sets; the screens have been removed, leaving the protruding backs of the old bulky sets (the generation before flat-screen TVs) as nifty planting pots filled with soil. It turns out that old computers, particularly old video display terminals rescued from the scrap heap, also make good vegetable beds. And these farmers are also tending an "alcohol garden," where old whiskey bottles—the Black & White and Imperial Blue brands—are the centerpieces of a novel drip irrigation system. The bottles are filled with water and quickly turned upside down and jammed into a trough of soil. The water slowly leaks into the soil, nourishing spinach, kale, and a variety of other vegetables that rise beside the bottles. The bottles also irrigate a small orchard of mango, guava, and banana trees.

Water and soil are precious commodities in the informal settlements. Although the Ngong and Nairobi rivers flow through and around the capital city, their waters are undrinkable, heavily polluted by agricultural runoff (topsoil and chemical fertilizers) from fields upstream, by human and industrial waste from factories and houses, and by seepage from the dump sites on the riverbanks. On a bright, sunny day, the river water looks pitch black, like a stream of oil. Still, women wash clothes, boys scrub cars, and children play in the water. Drinking water is so scarce that the municipality brings it in with tanker trucks a couple of times a week. The trucks squeeze down the narrow roads of the neighborhoods, blocking traffic. Hundreds of people line up with yellow plastic jerrycans and rush to the trucks to fill them when the water begins gushing from hoses.

One group of young urban farmers call themselves Komb-Green. "We comb youth away from crime and into the farm," explains one of their leaders. They have constructed a hydroponics system that doesn't need soil, but rather deploys ground-up coconut shells as peat-like planting material. An improvised irrigation network of plastic pipes and a pump channels rainwater from a collection tank into a series of wider pipes stuffed with the coconut shells. The seeds are planted in large nesting holes carved into the pipes. In full growth, the pipes are covered with the billowing leaves of spinach and lettuce.

Some of the produce is sold in the market to raise money for seeds and planting material. But, in the spirit of *Ubuntu*, most of the food grown by KombGreen and the other urban farmers of Nairobi is shared among the members who do the work and the wider community where the need is greatest. There is hot porridge for schoolchildren in the morning, and vegetables for families who have no food at night.

And eggs, thanks to the women of the Team Revolution garden group raising about two hundred chickens in wire coops tucked into an alleyway. The Cool Waters project provided the chickens and constructed the coops, and the women have taken it from there. "We as women decided to do poultry farming, so when these hens lay eggs we can sell the eggs for income. Eggs are so expensive now in Nairobi, it is a good market," says Rose Syotini, noting that the price of a single egg nearly doubled to 13 shillings (about 10 U.S. cents) during the pandemic. The money goes into a savings fund that the fifteen women in Team Revolution can tap to pay for school fees, medicine, and other emergency expenses. They also keep plenty of eggs at home for their own children to eat. Eggs, says Rose, have become central to their diets along with vegetables they grow around the coops. The chicken poop makes for good fertilizer.

"We are all mothers," says Rose, who has six children herself. "Our women are empowered to give our children protein and carbs and good nutrition." As for the name of their group? "We are leading a revolution, taking our food needs into our own hands," she says. "We must feed ourselves."

Greg also strikes a militant tone. As he approaches his City Shamba demonstration farm, he says, "Welcome to city solutions in the war against hunger and malnutrition."

The outside wall of City Shamba is a bright mural of a golden sun shining over cityscapes with green plants and patches of blue sky and water. It is perhaps the cheeriest wall in all of Nairobi. Behind it, in an abandoned, partially collapsed building, is a horticultural heaven. Vines and limbs from climbing plants and fruit trees—passion fruit, dragon fruit, avocados, tree tomatoes, beans—stretch overhead. The ground and walls are covered with all manner of planting containers hosting all manner of crops. Amaranth, cowpeas, onions, basil, spinach, chilies, kale, celery, parsley, dill. There are chickens and turtles and a terrarium with snails.

"People here don't eat snails, but our idea is to sell them to hotels to serve to people who do eat snails," says Greg. And, he notes, snail slime is in demand by the beauty industry for use in skin moisturizers. If he can find a way to harvest the slime, he would turn the added revenue into more vegetable seeds and tree seedlings.

In his garden, Greg has also planted herbs that produce scents that repel pests, eliminating the need for chemical pesticides. And he breeds black soldier flies in a separate screened-in room filled with flies, larvae, and maggots. The black soldier flies in their various life stages decompose organic waste collected from households and hotels and convert it into soil and compost, which is then used as fertilizer. The protein-rich larvae are also good feed for poultry and fish. Down the street, in a backyard clearing, Greg and friends are nurturing a tiny fish farm.

"We're purely organic," Greg says. "For a long time, Nairobi has been depending on the rural population to feed its residents. Those farmers use pesticides and chemical fertilizers. The quality can't be assured. If you buy in the markets, you don't know how safely the food is grown. Here we grow food that isn't compromising our health. We can take back the power to control what we are eating."

It is all part of his "grow your own" mantra. "Once you depend on others for food, you lose control of your food and your health. In the cities, people need to have money to purchase food. Maybe some days you don't have money, and that means for that day, you can't put food on the table for your family. But you can if you grow your own."

He created City Shamba to be a learning laboratory for urban farming and a proving ground for the possibilities. "We show how you can grow food wherever you want. On your balcony, your rooftop, in boxes or any containers. You don't need farmland. There are no limits to where you can grow food." He often welcomes school groups, hoping to inspire a youth-led urban agriculture lifestyle. He gives them seeds to take home and start growing on their own.

It is an opportunity he wishes he'd had in his school days. "Growing up as a young kid, there was no guarantee we could get three meals a day. I used to depend on the school meals," he says.

As Greg leaves City Shamba, a voice calls out from a rooftop garden across the street. "Thank you for the seedlings," a man shouts. "As you can see, I have used them."

Greg waves, and smiles. "You know what they say," he adds. "Every journey begins with one step."

For Cool Waters founder Elizabeth Kimani-Murage, her first step was in rural Kenya when her mother, a subsistence farmer, gave her a small plot of land. "I called it mine. I planted vegetables. It was very exciting," she says. Then she went off to school in the city, got her PhD, and began her research on the impact of urbanization.

Another mother jolted her back to her agriculture roots at the beginning of the Covid pandemic. Elizabeth recalls hearing a news story of a Kenyan mother boiling stones in a pot at night, creating the illusion of cooking food so her children could go to sleep with the hope there would be something to eat in the morning. "I thought, 'If only she had vegetables to cook instead of stones.'"

On Mother's Day 2020, as Nairobi's pandemic lockdown tightened, Elizabeth began turning her own house and yard into an edible landscape. It is in one of Nairobi's newest subdivisions, part of the inexorable dominion of concrete over land that once grew crops. Two years later, ripening strawberries form a red-and-green hem all around her house. Fruit trees line the driveway and the edge of her property. Apples, bananas, cherries, papayas, mangos, figs, pomegranates, tangerines. Vegetables carpet her backyard, flourish in a small greenhouse, sprout from a hydroponics experiment, and climb toward the sky in vertical plant beds. Kale, chilies, spinach, onions, cabbage, broccoli, celery, tomatoes, herbs, and spices. "We don't buy anymore, we plant," she says.

"Covid showed us that food sovereignty, growing your own, is more and more important. Most everybody has some space in their neighborhood to create a little food resilience," Elizabeth explains. "If as many people as possible in the city can do farming, we can create new urban micro-environments, with a cooling effect from trees and vegetation, cleaner air."

As she looks to 2050, she winces at the prospect of a Nairobi with fifteen million inhabitants, making it one of the most-populous metropolises in the world. But then she closes her eyes and sees vegetables everywhere, on balconies, on roadsides, between houses, in schoolyards. She sees canopies of fruit trees nourishing and cooling the city. "And I see people sharing food in the spirit of *Ubuntu*. And there is no one hungry."

She knows it may sound fancifully utopian. But that is her vision for nourishing the community, preserving the environment, and restoring the city to the original inspiration of the Maasai, a place of cool waters.

"Nairobi is our starting place," Elizabeth says. "We will go from here."

As she speaks, across the Indian Ocean, in an even more populous, densely crowded city, a similar inspiration stirs.

CHAPTER 4

THE INDO-GANGETIC PLAIN

When Will We Learn?

THE NOVEMBER AIR DRIFTING south from the Himalayas sits heavy on Delhi, the world's second-most populous urban area. The blanket of smog is heavier and more polluted than usual, for November is the time of year when brush fires rage on the fields of the Indian subcontinent's breadbasket, encasing the vast landscape in a gray, smothering, smoky haze. The farmers north of Delhi are burning the stubble from their freshly harvested rice fields so they can quickly convert them to wheat fields. Time is of the essence in the churning crop cycles of these two major grains, as it is in all agricultural endeavors in a country of more than 1.4 billion people with malnutrition rates among the highest in the world. Satellites track the smoke

Left: *Green Evolution*
Urban smallholder farmers harvesting vegetables and herbs in Delhi

plumes rising from the farms as they spread greenhouse gases and particulate pollution over the northern part of the country. Newspapers and television stations chart the impact on the Air Quality Index. "Farm Fires Drag AQI Into Very Poor Zone—Punjab Stubble Burning Count at Season High," reported a headline in the *Times of India* in November 2022. The paper noted that satellites detected more than four thousand fires in the states of Punjab, Haryana, and Uttar Pradesh, four hundred more than the previous daily record of the season.

Farmer Surendra Singh, tilling a couple of acres of land on the edge of Delhi, curses the leaden skies above. "When will we learn?" he asks, coughing weakly. He grows spinach, mustard, beets, cabbage, sweet corn, and fenugreek, a staple spice and herb of Indian cuisine. "Look what the smog is doing," he says, pointing to white spots on his spinach leaves that he fears will reduce their value in the market. A neighboring farmer, Satish Kumar, joins him with worries about wilting cabbage leaves and the tepid color that portends possible stunting of his bottle gourds, a favorite ingredient in soups and curries. "It's the healthiest food there is," he says. In unison, the two men blame other farmers for this damage to their crops, for torching the rice stubble and creating the smog that deprives their vegetables of needed sunlight and traps the withering heat in place. It is a classic example of agriculture turning on itself.

Out of the smog emerges the determination of these smallholder farmers to confront the Great Collision in one of the world's oldest farming zones. They have enlisted in a post–Green Revolution revolution, to heal the air and the land with healthier crops—and they want all other farmers to join them, especially their stubble-burning neighbors. This stretch of vegetables on the outskirts of the largest city in the world's most-populous country is their starting place.

"We can't go on this way," Surendra says.

"Something has to change," Satish agrees.

"We know there is a better way to do things," Surendra insists. "There has to be."

◈

The Indo-Gangetic Plain sweeps beneath the Himalayas, across eastern Pakistan, through northern India, and into Bangladesh, stretching from the Thar Desert in the west to the Bay of Bengal in the east. Blessed with rich alluvial soils from the Ganges, Indus, and Brahmaputra rivers and mountain tributaries that deposited mineral- and nutrient-laced sediment through the centuries, the Plain is the world's most-farmed region. Farmers here traditionally practiced subsistence agriculture, growing to feed their families, but transitioned to more commercial, profit-oriented enterprises during British colonial rule. Then the Green Revolution arrived in the 1960s, with hardy, fast-growing wheat strains and improved rice varieties, and the Indo-Gangetic Plain soon became an agricultural powerhouse. Out of famine, the country ascended to the world's second-largest producer of wheat and rice. Grain grew almost everywhere—between houses and schools, around factories and gas stations. Harvests were so huge that schools were closed and converted into grain warehouses. Wagons groaning under huge mounds of wheat choked the roads. Farmers stored surplus grain in their homes; wheat spilled out of front doors and bedroom windows. By the turn of the century, the national stockpile of wheat and rice was approaching sixty million tons. The government funneled much of the harvest into a massive public food distribution system intended to feed the poor. As the twenty-first century progressed, India's annual wheat production surpassed one hundred million tons and it became a global exporter.

All the while, the population was also dramatically expanding. From the time of the Indus Valley Civilization in the Bronze Age (one of the world's earliest civilizations, along with Egypt and Mesopotamia), the fertile land of the Indo-Gangetic Plain has given rise to great urban centers. It is India's most densely populated area and home to nearly 10 percent of the planet's population. At the time Surendra Singh was inspecting his spinach crop, India was overtaking China as the world's most-populous country. Delhi, with about thirty-two million people, has an annual growth rate exceeding 2 percent from new births and migration into the city; its population is projected to hit forty million by 2035. Which would put it on pace to become the world's largest urban area.

Here, on the Indo-Gangetic Plain, the Great Collision looks particularly ominous. The increasing population ratchets up the pressure on agriculture to keep expanding production, which puts pressure on the health of the people and the environment. The Green Revolution's food production boost depended on increased use of chemical fertilizer, pesticides, and herbicides, readily flowing irrigation, and government subsidies to pay for it all. But rather than scaling back on these expensive and environmentally harmful practices over time, Indian farmers became more dependent on them. And so, with little respite, water resources continued to be tapped to flood and irrigate the fields, making India the world's largest user of groundwater. Chemicals were poured on the crops and into the soils in increasing amounts to spur growth and protect from pests and diseases. The rice-wheat cropping cycles on the world's largest alluvial plain spun ever faster: rice planted in May and harvested in October–November, wheat planted immediately afterward in November and harvested in April–May, followed again by rice. Burning the crop residue left after harvesting was popularized as a way for farmers to hastily clear their fields so the next crop could be planted; cutting the stubble and mixing its nutrients back into the soil was considered to be more time-consuming, a luxury that could delay the next planting and thereby derail the entire seasonal crop cycle.

The planting seasons of 2021–22 profoundly exposed the effect of changing climate patterns on agricultural production. The late onset of the 2021 summer monsoon rains caused delays in the sowing of rice. This pushed the rice harvest back a couple of weeks, which in turn delayed the wheat planting by a similar length of time, increasing the risk of late-season heat stress on the wheat crop in the following March and April. That risk became reality. An unprecedented heat wave arrived in mid-to-late March 2022, pushing average temperatures for that month and April to highs not seen over the past hundred years. The record heat, well past 100 degrees for days on end, torched the ripening wheat at a relatively early stage, causing considerable damage. The knock-on effect was clear: if the monsoon rains hadn't arrived late, the wheat would have been more mature and closer to harvest when the heat wave hit the following season.

This domino effect in the fields of India—one late rainy season disrupting the agricultural seasons to follow—rattled the global food chain. The rice-wheat rotations in India and all of South Asia account for nearly one-quarter of the world's food production. The Indo-Gangetic Plain feeds 40 percent of the Indian population. The two largest wheat- and rice-producing states, Punjab and Haryana, contribute almost 30 percent of India's total wheat production and supply over 60 percent of the government grain buffer stocks. So imagine the impact when those two states suffered a wheat yield loss upwards of 15 percent in 2022, according to estimates of the International Maize and Wheat Improvement Center (CIMMYT). A follow-up modeling simulation in 2023 by CIMMYT predicted that climate extremes magnifying heat and drought stress would trigger average declines in wheat yields of 16 percent in South Asia and 15 percent in Africa by 2050.

The Intergovernmental Panel on Climate Change, which won the Nobel Peace Prize in 2007 for sounding the alarm on the impact of global warming, lists India as among the countries expected to be most affected by the climate crisis. In 2023, as unbearable heat once again bore down on the Indo-Gangetic Plain, and the rains were as erratic as ever—too early, too late, too little, too much—the University of Cambridge in England issued a study saying heat waves in India were putting "unprecedented burdens" on the country's agriculture and economy. It warned: "Long-term projections indicate that Indian heatwaves could cross the survivability limit for a healthy human resting in the shade by 2050."

India in particular has become a roiling cauldron of the leading agents of climate change. It is home to more cows, which have been central to Indian civilization for centuries, than any other country in the world. It is the world's second-largest rice producer. It is also the world's second-largest producer of food waste; about 40 percent of all food grown in India is lost along the supply chain because of poor post-harvest handling or is discarded by consumers (Delhi's largest landfill is, at 240 feet, about as tall as the Taj Mahal, and growing by the day). All three of these factors—belching cows, decomposing rice paddies, landfills stuffed with rotting food—are the major sources of methane. India is also the world's second-largest user of nitrogen fertilizer. All of this contributes to making India the world's second-largest emitter of methane and the third-largest emitter of all greenhouse gases. Those are the very same gases that trap the sun's heat in the atmosphere and lead to longer and more intense heat waves, like the one that smothered India's wheat crop in 2022. In India, agriculture is just behind the energy sector as the largest source of these emissions.

The Indo-Gangetic Plain thus becomes a prime exhibit in the case of Agriculture vs. Agriculture. Here, the farming that delivered many millions of people from famine has placed the task of growing food for future generations in greater jeopardy than ever.

And through it all, despite the relentless push to produce, produce, produce over the past half century, malnutrition and micronutrient deficiency stubbornly persist, particularly in farming families. India is home to one-third of all the malnourished children in the world. It still ranks near the bottom of the Global Hunger Index. More than 30 percent of its own children under five years of age are stunted in some manner—physically, cognitively, or both—from malnutrition in their formative early years.

◈

The post–Green Revolution revolution—with Surendra Singh and Satish Kumar as pioneering foot soldiers—aims to not just *feed* the growing population with calories but to *nourish* it as well. "At the time of the Green Revolution, we needed production. The priority was food for energy—calories, not nutrition," says Vikash Abraham, as he leads the way around Delhi's Regenerative Agriculture Cluster established by India's Naandi Foundation. "Our focus is on nutrition-dense crops."

At his feet are row after row of crops that are decidedly not the energy- and calorie-packed wheat and rice that blanket much of the Plain, but rather an array of nutrition-packed crops:

kohlrabi, parsley, green peas, fenugreek, broccoli, red cabbage, red lettuce, kale, coriander, carrots, turmeric, dill, spinach, beet root, Swiss chard, amaranth, cauliflower, zucchini. It is a cornucopia of possibilities.

Naandi calls its model Arakunomics, which seeks a "yes" to the following questions: For every food that you eat, can you ensure that the farmer who produced it is profitable? Can you ensure it's good for you? Can you ensure that it's not at a cost to the planet? It is both a practice and philosophy that emerged from Naandi's two decades of work with smallholder farmers, particularly in the tribal region of the Araku Valley, where it lifted destitute farmers out of extreme poverty with a new coffee-growing culture. During that time, farmers across India, especially those working only an acre or two to feed their families, were experiencing sharp declines in soil fertility precipitated by relentless monocropping and an over-reliance on chemical fertilizer. They watched these chemicals run off their fields and dirty their water, and they noticed their underground water tables shrinking from free-flowing irrigation practices. They puzzled over weather patterns becoming more unpredictable and extreme, and they breathed ever-more polluted air. They saw pests and diseases defying the insecticides and herbicides they generously sprayed on their crops at great expense and risk to their health. And they watched prices for these inputs and seeds go up and up while their profits went down, often below zero. Debts piled up. No matter how diligently they worked their land, malnutrition still threatened their children. Despair spread. Suicides became tragically routine in the countryside—an average of more than thirty a day, according to government reports in 2020.

Agriculture, at least for the smallholders, had gone terribly wrong. Arakunomics, Naandi believed, would reverse this with a regenerative agriculture system "where farming is once again profitable and environmentally sound, and where healthy food is affordable and bountiful," explains Vikash, the foundation's chief strategy officer. It may sound as utopian as the vision of Cool Waters Nairobi, but as the farmers of the Great Rift Valley are discovering, regeneration is the only answer to their desperation. Naandi's vision, like Cool Waters', attracted the support of the Rockefeller Foundation, which had been a major supporter of Norman Borlaug's work. Now, here on the peri-urban fringe of Delhi, on Naandi's 250-acre cluster of organic farmers, the Green Revolution comes full circle.

As a first step toward engaging smallholder farmers in this transition from Green Revolution to Evergreen Evolution, Naandi established Urban Farms Co. in 2019. In this model, farmers on their small family plots are weaned off chemical fertilizer and pesticides and seasonal monocropping. Instead of laying down a solid blanket of wheat or rice, they now tend to a quilt of three or four different crops each season. Rather than growing for surplus production to fill a strategic reserve, they now grow for the market, nurturing crops in high demand by the nearby urban consumers who have largely been dependent on long-distance supply chains often disrupted by economic breakdowns and plagued by immense food waste during transport. The farmers receive their seeds and organic compost fertilizer from the cluster's service hub, which guarantees to purchase their crops at a 20–30 percent premium price for organic produce and then distribute it to retail markets. Following Delhi, Urban Farms planted additional clusters in the massive urban centers of Bangalore, Hyderabad, and Mumbai. Within three years, 150,000 farmers were growing in these regenerative agriculture clusters, facilitated by ten service hubs. When farmers bring their produce to the hubs, they are greeted by a poster proclaiming, "The Future of Farming Is: Soil Biodiversity. Carbon Sequestration. Water Retention."

For farmers who have depended on government-subsidized chemical fertilizers, pesticides, and herbicides for decades, this new organic, regenerative business was a scary proposition. "People feel they don't need to change, they don't want to change," even if farm losses keep piling up, Vikash says. Government attempts to alter the hidebound subsidies regime and price supports that began during the Green Revolution have been met by nationwide farmer protests. Urban Farms, with its even more radical proposition, faced a mountain of skepticism.

"For farmers who get their livelihood from just one or two acres, to transition away from conventional farming to regenerative farming can be seen as a big risk," Vikash explains. "They can't afford to let an acre be less productive in the first year of a transition. So from day one that acre has to generate the same profit. If you have one hundred acres, you can afford to set aside one or two acres for a trial. We had to do things so farmers could see improvements right away. People won't tolerate longer transition periods."

Moving farmers off the land to let it heal, as was done in Ethiopia, wouldn't work here, Naandi calculated. A new system of support was needed. "So," Vikash continues, "we set up the service hubs, a one-stop place, very convenient to the farmers to adjust to changes. We have economies of scale, we reduce wastage. We have a market that gives value to the produce."

The first challenge was to break the farmers' addiction to chemical fertilizers. Naandi knew they needed to find a way to make large amounts of organic compost and deliver it to farmers on time. "If you aren't providing chemicals for crops, you have to provide an alternative for plant nutrition," Vikash says. So Urban Farms turned two of India's biggest agricultural and environmental challenges—the rice and wheat stubble destined for burning and the abundance of dung from the nation's cows—into assets. Both became prized ingredients in the homemade organic fertilizer. The nutrient-packed stubble is collected directly from fields, relieving farmers from having to burn it, and the nitrogen-rich dung is purchased from *gaushalas*, the traditional shelters that care for old and infirm cows. Urban Farms workers also clear public lands of weeds and wild brush growth and rescue discarded plant greens from ending up in landfills, including discarded banana leaves from a banana ripening center. All this biomass is mixed together along with various micronutrients and microbes that help enrich and speed up the composting. Each hub, servicing about 250 acres, produces more than two thousand tons of compost annually.

At the service hub in Delhi, a thirteen-acre complex about a one-hour drive from the city center, eight long rows of thick compost mounds, each several feet tall, ferment in the sun. About seven tons of cow dung arrive every couple of days from the *gaushalas*. A team of workers shape the mounds, layering the dung over the stubble and greens. Wielding pitchforks and maneuvering bulldozers, they aerate the soil and regulate the internal temperature. After ninety days of composting, the mixture is ready to be distributed. "All a farmer needs to do is call us and tell us how much they need," says Vikash.

In a series of demonstration plots, where seeds are buried in soil rich with the compost, the potential of Naandi's regenerative vision comes to life. More than forty varieties of lush crops beckon the farmers. In the first season, curious farmers came to look, but none signed up. Yes, they agreed, the flourishing fenugreek and coriander and parsley looked enticing. But who knew how to grow those crops? Was there a market for them? Where was the wheat and rice and maize? And how do you make things grow without chemical fertilizers? Surely, they suspected, Urban Farms was secretly adding fertilizer to the soil at night, under the cover of dark.

It was a radish panic that helped win over an initial batch of farmers. A group of local farmers was famous for growing tasty radishes, but just when the Delhi cluster was beginning, a fungal disease arrived, discoloring their radishes with unsightly spots. The markets were rejecting the tarnished radishes, rendering them worthless, and the farmers had no option but to plow back into the ground what their families couldn't eat. They lost a season of income. Debts mounted. Some of the farmers, out of desperation, took their spotted radishes to this new enterprise called Urban Farms that promised to revolutionize farming. "We didn't know about radishes," Vikash says. But Naandi did know that the farmers blanketed their fields with heavy doses of chemicals. Urban Farms planted radishes in their organic compost with no spraying. After forty days, they harvested spotless radishes. The farmers were astonished. One told Vikash, "What you guys are doing seems to work. You seem to be on to something." The radish farmers were among the first to sign up for the Urban Farms program. And then others followed. Soon, there were more than two hundred working with the Delhi hub.

"They joined initially to make more money, but after a year they see their soil is better, and other benefits," Vikash says. "Regenerative agriculture is about the biology, the more nutrients you have in the soil. It is about practices like crop diversity, staggered planting. If small farmers can be given advantage of scale, they become artisan producers. The attention that a family can give to one acre of land is more than a larger farmer can give to all his land."

Urban Farms also integrated natural pest management into its demonstration plots. Interspersed with the vegetables, herbs, and spices are plants, some of them weeds, that repel pests with their odors. Others, known as trap plants, have scents that attract pests so they stay away from the crops. Bright yellow-orange marigolds, ever-present at Indian festivals and weddings, grow at the ends of many rows. For all their beauty, marigolds have a scent that keeps away pests like mosquitoes and nematodes and attracts other insects that attack and kill aphids. All this plant-life seduction and repulsion relieves the farmers of the need to spray toxic pesticides and herbicides.

At the hub on the edge of Delhi, one of four in northern India, farmers arrive with wagonloads of freshly harvested produce throughout the day, moving directly and swiftly from vine, tree, or field to the central collection point. There, the crops are weighed, cleaned, sorted, graded for quality, and kept in refrigerated storage. As night descends, the produce is transported to a warehouse near the Delhi city markets and repackaged for retail sale. Before dawn, trucks deliver the produce to a network of seventy stores and e-commerce distribution sites. The first season of production was in 2020, during the Covid pandemic shutdown, so deliveries went straight to consumers in their homes. "While everyone was stopped, we were starting," Vikash says. By its third year of operation, the Delhi site was churning through ten to twelve tons of produce every day, from farmers to consumers. This distribution network overcomes three additional challenges of Indian agriculture: lack of on-farm refrigerated storage, long-distance supply chains, and stale product in the markets—all of which result in substantial food waste, tremendous income loss for the farmers, and squandered potential nutrition for consumers.

Choreographing all of this is an Urban Farms sales team

that monitors markets and seeks to time farmer harvests with consumer demand for specific vegetables. During the Green Revolution's single-minded push to increase production, there was little consideration of markets.

"Timing production with the market is rarely done in India. It is difficult for farmers to do themselves. Someone needed to take responsibility for the entire chain," Vikash says. "Every day, vegetables are needed for the market, and certain vegetables during certain times, so we have to stagger production by the farmers to match demand. Once farmers see someone else making money from tomatoes, for instance, then everyone starts growing tomatoes and the price falls with the surplus." And the farmers suffer.

Satish Kumar was one of the first farmers to work with Urban Farms. He approached cautiously, not going all-in at first, but testing the regenerative model with one-half acre. "I could see the opportunities and wanted to move to be more organic. I was thinking my land was declining over time from an overuse of chemicals. And the fertilizer costs were going up," he says while inspecting his green bottle gourds. They were taking on a silver sheen as the smog created by other farmers settled over his land. "And I wanted to grow vegetables that could provide more nourishment."

That first season of growing vegetables encouraged him to devote his other four acres to Urban Farms as well. With the higher prices for the organic produce, his income doubled. And his costs declined, particularly since he no longer had to pay middlemen for transporting his crops to markets. In the November season, he and his wife, along with their two children, tended to patches of cabbage, coriander, and potatoes beside the bottle gourds. In other seasons, he grows beets, sweet corn, and pumpkins. He also has a plot of rice, and cherishes the stubble for composting. He has learned it is too valuable for it to go up in smoke. He also knows that his neighboring farmers are paying close attention to his actions, pondering joining Urban Farms themselves. "They come by to see how I'm doing," he says.

One of them was Tilak Raj, who kept an eye on Satish's fields as the vegetables flourished. "I could see that he was getting regular visits from the field staff, and I was impressed with the advice," he says. "Nobody had ever come and observed and listened to our problems. Nobody." His acres sit right on the Delhi city limit, bordering on the state of Haryana. For ten years he worked here, and no one from either the Delhi or Haryana governments had ever mentioned the impact of chemicals and monocropping on his soils. After watching the delivery of compost to Satish, Tilak joined Urban Farms with three acres.

He farms barefoot, cutting fenugreek for a nighttime delivery to the hub. Several family members with machetes trim the bottom stems to meet market preferences. In the coming days, they will also harvest cauliflower, coriander, beets, gourds, and spinach. Tilak is grateful for the higher prices that this organic produce fetches in the markets, and expects more in the future as demand grows. He explains with this analogy: "If you have one pair of shoes for 100 rupees and one pair for 5,000, you expect that the expensive shoes will be of better quality. With food there should be better prices for better food."

Vikash and the other farmers laugh and nod their agreement. But the laughter stops when Vikash tells them about the soaring prices for the chemical fertilizer they formerly used. The war in Ukraine, he explains, caused a shortage of many components, including urea. While the farmers working with

Urban Farms get plenty of urea from the cow dung and urine provided by the *gaushalas* and mixed in with their compost, the vast majority of farmers in India who still depend on the manufactured versions are facing grave shortages, Vikash reports. Farmer Surendra Singh, who cursed the stubble-burning smog tainting his spinach, has no sympathy for those still using the chemicals.

"That's good," he says of the chemical fertilizer shortage. "They should stop sales altogether. We know there's better ways to farm. We're proving that."

◈

Further north on the Indo-Gangetic Plain, in the state of Punjab, the source of much of the stubble-burning smoke, orchards blossom where wheat and cotton once flourished during the Green Revolution. A group of a half-dozen farmers reverently admire the fruit of their labors: a kinnow, a juicy type of mandarin orange.

Surinder Charaya, one of the leading growers, shows off a bowl full of the bright citrus fruit, holding it aloft like a hard-earned trophy. It is a remarkable accomplishment, he says. Not just that he has a bowl of kinnows on his desk, for growing them is his business. But that he has a bowl of kinnows on his desk *now*, four months after harvest. That, he says, is a minor miracle.

"Several years ago, we wouldn't be eating kinnows at this time," he explains. "They all would be spoiled by now."

He peels one himself, juice squirting into the air. Then he passes the bowl off to the other farmers and offers a kinnow to me. Feel how cold it is, he says. See how shiny it looks. Now everybody is eating kinnows.

"So fresh," adds a second farmer.

"Still succulent," marvels a third.

Surinder motions to another room beyond his office at Balaji Kinnow Co. "Cold storage," he says. It is what keeps their fruit fresh for months after the harvest. This room was added a year earlier, the final link in a new cold chain. The farmers agree: it is "revolutionary."

In the orchards around the town of Abohar, about 250 miles north of Delhi, yet another agriculture revolution is spreading. After the Green Revolution and the White Revolution (also known as Operation Flood, which propelled India to become the world's leading milk producer), the Orange Revolution dawned. This one—also known as the Cool Revolution—aimed at reducing the staggering amount of food loss and waste of Indian agriculture. After boosting harvests by multitudes during the Green Revolution, the focus here has shifted to preserving all that bountiful production while making sure the food, and the nutrients, actually reach consumers.

Before the adoption of cold chain technology, which preserved Surinder's bowl of fruit, the kinnow farmers would routinely lose 20 percent of their crop in the immediate post-harvest days, failing to quickly cool the fruit for storage beyond a week. Another third would be lost in transport, spoiling in bags and boxes on the backs of open trucks. The more distant the market, the greater the loss. During its agriculture boom, India became the world's second-largest fruit producer (behind China). But it also became the world's leading fruit waster, with up to half of that production spoiling from tree to market.

"We've had a flood of food thanks to the Green Revolution, but we needed to ask, 'How do we convert from production to satisfying demand?' We now needed to focus on preserving all that food," says Pawanexh Kohli, the head of India's National Center for Cold-Chain Development. "Can you imagine any industry, say cars, where 30 percent is wasted and you still ramp up production? Wouldn't you first figure out a way to end the waste?"

No part of the world was impacted more profoundly by the Green Revolution than India's Punjab region. It was a land of hardy grains, oilseed crops, and cotton when kinnow was introduced in the 1930s by researchers at the University of California. It was the one place in the world where the fruit flourished, on both sides of what became the India-Pakistan border. Then Borlaug's wheat arrived in the 1960s, and this new golden crop thrived across the Punjabi plains with the aid of fertilizer and new science to combat disease and pests. As more water resources from the Himalayas flowed through irrigation canals, rice was introduced, even though it wasn't a main part of the local diet. Farmers became wealthier with the thriving wheat and rice. During the grain boom, the kinnow faded in importance. Its development was dormant for three decades.

In the 1990s, a kinnow renaissance emerged as the agriculture environment changed. Boll weevils wiped out the cotton crop. Soils began to deteriorate from the wheat and rice dominance and demanded more and more chemical fertilizer at ever greater expense. Underground water tables dipped drastically; wells that once filled with water at twenty feet now were dry to a depth of two hundred feet. Farmers began shifting their investments back to their kinnow orchards, which required less fertilizer and a more judicious use of water. They discovered that wheat thrived beside the kinnow saplings as the trees grew to full height. Gradually, the blanket of golden grain that once covered the landscape around Abohar became a quilt of gold and green—wheat amid the kinnow trees. The region that once had more grain than it knew what to do with became famous for its fruit.

The farmers gathered at the Balaji office were at the vanguard of this shift, and were eager to lead a tour of their orchards. Most were twenty-five to forty acres, with about one hundred trees per acre, each tree growing to about ten feet tall. A good harvest would be 1,500 kinnows per tree; in a bumper harvest, the branches would bend with 4,000. Water, the farmers said, is particularly precious in the kinnow fields. The groundwater had a high saline content; after decades of steep fertilizer use, the streams and rivers were laden with chemical runoff and pollutants from the swelling urban and industrial areas. The farmers constructed a large retention tank that would fill with water from the canals running down from the mountains and the annual monsoon rains. A small control room regulated the flow of water to irrigate the orchards at prescribed intervals depending on the weather, and to mix fertilizer with the water to nourish the soil. The old method of flood irrigation, similar to the practice that saturated the rice fields, could trigger fungal diseases that caused the fruit to drop prematurely. To end this waste, the farmers began switching to automated drip irrigation, which keeps the fruit on the trees longer so the peels thicken a bit more, bestowing extra shelf life after harvest.

As kinnow production doubled in a five-year period and demand for the juicy fruit rose across the country, so did a problem all too familiar to Indian agriculture: storage and handling. The same farmers who were once overwhelmed by their bumper grain production now struggled with a fruit surplus. Initially, they stored the kinnows in gunnysacks, but the burlap material smothered the fruit and rotting set in quickly. For transport, they stacked the kinnows several rows high inside boxes. The boxes were stacked, too, one on top of the other. Then the journey commenced, often in temperatures nearing

100 degrees, over fairly wretched roads, for a couple of days, depending on the distance. Upon arrival at the markets, the kinnows in the bottom rows of the bottom boxes were bruised and rotting, the promised succulence vanished. The farmers' reputation among the vendors, who promised to deliver freshness to consumers, was also spoiled.

The farmers of Balaji Kinnow were desperate to reduce the waste and polish their image. Satish Setia, who had been nurturing kinnow trees since he was a young man in the 1960s, often marveled at how nature beautifully engineered the kinnow, with each fruit divided into ten identical slices snugly encased in a shell of orange peel. "Nature created a perfect air-tight packaging for the slices of fruit," he would say, "and man can't develop a packaging system to prevent wasting it after harvest?"

The farmers approached the government and the agriculture institutes for advice, but they found little support. The government was still fixated on wheat and rice and the rising costs of subsidies to prop up grain production; it didn't need another group of farmers clamoring for assistance. Instead, farmers across the country were being told they needed to reduce their dependence on public support. And besides, the government and the scientific community were setting their sights on a new horizon: space exploration, with a mission to the moon and beyond.

So the farmers of Balaji concocted their own solutions. In 2006, they installed new packing technology to wash and grade the quality of the kinnows after harvest, and to wrap them in a wax that seals in the freshness, prevents mold and fungus from developing, and extends the shelf life by a couple of weeks. They then pioneered the use of a pre-cooling system designed for the rapid removal of heat from freshly harvested produce, which prevents over-ripening. And they added cold storage facilities to preserve the fruit until shipping. All those advances, they found, reduced waste by about 25 percent.

But transport remained a problem. Continued increases in production saturated the local markets and the farmers were in search of more distant consumers. Balaji designed plastic crates that allowed greater air circulation among the kinnows and added buffers that would cushion against bruising during transport. Still, the long drives to the big cities in the south of the country, thousands of miles away, took a toll with spoilage rates of about 25 to 30 percent. That could be more than twenty thousand kinnows per truck, an especially tragic loss in a country with such high child malnutrition rates.

Refrigerated trucks, common in other parts of the world—including China, India's main rival in fruit growing—were still a novelty in India. You could drive for hours on the paved highways and dirt back roads and encounter all sorts of conveyances hauling all manner of crops, but nary a refrigerated vehicle. Again, the frontline push across all agriculture was perpetually larger harvests; post-harvest concerns were a nuisance. The Balaji farmers, though, were tantalized by promises that these refrigerated trucks could nearly eliminate waste, which seemed impossible. Until they participated in a trial run.

In the trial, conducted by the Indian School of Business, the National Center for Cold-Chain Development, and Carrier, the international refrigeration company, a refrigerated truck set off from Abohar filled with Balaji kinnows on its way to Bangalore in southern India, an overland journey of more than 1,500 miles and five days. When the truck arrived, less than 10 percent of the fruit was damaged, rather than the 30 percent transport waste the farmers had built into their costs. They couldn't believe their eyes. In total, the entire cold chain from post-harvest handling to cold storage to refrigerated trucks had reduced their waste by nearly 80 percent.

That meant more kinnows to sell. Many more. The farmers quickly calculated the increased income from the reduced waste. It came to about 150,000 rupees (nearly $2,000 then) per truck. With a fleet of refrigerated trucks, they imagined they could increase their overall income tenfold. Surinder Charaya, the owner of Balaji, planned to deploy hundreds, then thousands of refrigerated trucks. He had been working in the orchards since he was a boy, selling kinnows from a bike. He left high school to grow the fruit business and get kinnows into everyone's hands. Now, distance was no longer a barrier to his ambitions. The kinnows could survive any journey. He began plotting routes to all the major urban centers of India, and even exports to neighboring Bangladesh and beyond, to Dubai and Europe.

"The cold chain is a bridge to the markets, connecting supply to demand. The minute you connect to markets, the farmers become more productive. We saw how the wheat and rice farmers improved production because the government was buying their wheat and rice," explained Pawanexh Kohli, who grew up in Punjab during the Green Revolution and went on to lead the Cool Revolution. "The Green Revolution has come and gone, the whole mindset that we need to produce more intensively or we'll all die is gone. Okay, there was a time for that. But we have to stop flogging a Green Revolution 2.0 or 3 or 4. We'll be destroying Mother Earth. Now we produce enough. We need to improve the food handling system of what we produce. We lose 100 percent of what we are incapable of handling. Nothing from Mother Earth should go to waste."

Now others came to Abohar to sample the kinnows months beyond the harvest and to seek the farmers' advice. "People who should be teaching others are now coming here to learn from us," says Pardeep Dawra, one of the farmers gathered at the Balaji office. Laughter fills the room. As does ambition. There is talk of seedless kinnows, of a range of juices, of expanding to other fruits to fill up the cold chain facilities when kinnows are still ripening on the trees.

Kohli and the farmers recognize that there is an environmental cost to refrigeration, with the release of carbon compounds that add to global warming. But they maintain it is less than the amount of greenhouse gases released by mountains of rotting fruit. And how do you calculate the value of the desperately needed nutrition lost to waste? They respect the impact of changing climates and warming temperatures; they see it in their orchards. The heat wave of 2022 hit their trees at a critical stage of flowering, interrupting fruit growth. With continuing water shortages and erratic monsoons, some farmers recorded production drops as large as 25 percent.

But they believe the Cool Revolution has given them boundless opportunity, even if it did arrive late. "You know, the cold chain is an Indian concept," Kohli says with a wink. "Thousands of years ago, the meditation people would go to the Himalayas to absorb the cold air, slow their metabolism, so they could sit and meditate for days and weeks at a time."

As he sits in his air-conditioned office in sweltering Delhi, this is what he now contemplates: by reducing food loss and waste through better handling, India can be a model for easing the pressure of relentless production. After decades of intensified farming, he suggests, India can now be a country that releases land from agriculture and returns it to nature. It can lead the way in nourishing people and the planet.

That would really be something, he imagines. A cool journey indeed: from wasting to preserving.

◈

South of Delhi, the city of Agra is famous around the world as the home of the Taj Mahal, the immense and grandiose white marble mausoleum built between 1631 and 1648 by Mughal emperor Shah Jahan in memory of his wife Mumtaz Mahal. It is hailed as one of the new seven wonders of the world and stands as a monument to eternal love.

In India, Agra is also famous for its potatoes. Since Tikam Singh planted potatoes as an experiment in the 1930s, the Agra district has been the center of Indian potato production. But maybe not for much longer.

"We have been the largest potato growing belt in all of India. The most profitable potato is right here," says Yashpal Singh, Tikam's grandson, whose family now tends two hundred acres of potatoes. But in recent years, a neighboring region just to the north has had a bigger crop. "We're losing our place," Yashpal says.

Yields on his land have been slipping in recent years, and black spots started appearing on Yashpal's potatoes and those of his neighbors. He feared a case of a fungal disease called black scurf. Or, even more worrisome, was it a sign that potato blight, which has historically plagued potatoes in other parts of the world and other regions of India, had arrived here? He runs the TS Cold Storage facility, where his potatoes and those of neighboring farmers fill six levels of a refrigerated warehouse. The Cool Revolution has also spurred a proliferation of several hundred cold storage centers throughout the Agra potato belt. At his warehouse, Yashpal closely monitors and grades the quality of the potatoes coming in from nearby farms. He calculates that about 5 percent have the dark spots, but notes that the frequency is increasing as the disease spreads from farm to farm.

The thought makes him shiver. “What if it increases to 50 percent?” he says with alarm. Already, the market has knocked down the price of the spotted potatoes by 20 percent. The flesh of the potato still looks good, Yashpal insists, but consumers, ever more discerning, don’t like the dark spots.

He isn’t certain who or what to blame, though he suspects climate change has something to do with it. One of the causes of potato blight is warm, wet weather, which has been more frequent here because of heat waves and erratic rains. A spell of heavy rain and high heat and humidity in October of 2022, after the normal monsoon period, drenched the Agra fields at a crucial stage of potato planting. He says farmers also noticed the appearance of a white grub during those particularly wet periods.

Yashpal believes the farming practices over the years have taken a toll on the soil. His grandfather, Tikam, first planted potatoes here ninety years ago. A relative at a science research institute suggested he experiment with several crops like sugar cane, cauliflower, and potatoes, but the potatoes did the best. Since then, the family has done likewise, monocropping by planting only potatoes each year.

With the appearance of the dark spots, an agricultural advisor suggested to Yashpal that his potatoes might do better if he rotated crops, mixing in a variety of vegetables or mustard seed. That might prevent buildup of the disease in the soil and allow the land to restore from season to season. Indeed, Yashpal had heard that the rival potato farmers up north, who had thus far escaped the black spots, had begun diversifying their fields, planting potatoes along with cabbage and cauliflower. But Yashpal was skeptical. Although potatoes and cauliflower are often matched together in Indian cuisine—aloo gobi is a popular dish—he couldn't imagine mixing them together in his fields. Besides, he reasoned, all his equipment is tailored for planting and harvesting potatoes, and his cold storage center is designed for potatoes. And water scarcity limited them to one planting season per year. Would he risk that precious window by trying something new?

"We plant potatoes," he says. "That is our business. It's what we do."

His solution, instead, was to reject another orthodoxy of the Green Revolution and change the way they raise potatoes. He would go organic. His grandfather never used chemical fertilizer or pesticides. It was only in the past decade or so that Yashpal had increasingly relied on chemical assistance to reverse falling yields. Now he wondered if that might be the culprit. Were these actions of his to increase production now making that endeavor more difficult? He feared the chemicals had been killing natural organisms in the soil that were beneficial to his potatoes. He also suspected that the equipment used to turn the soil was too violent. Had the years of discing his fields to prepare for planting killed the earthworms and other insects? "We need those worms. They are really important for us," he says.

So he is going lighter on the tilling, disturbing the soil less. He is relying more on cow dung for fertilizer and depending on natural compounds from other plants for pest and disease control, like the Urban Farms model. His potato yield went down in the first year of the transition but rebounded the second season. After a couple more years of chemical-free farming, he can apply for organic certification, which will fetch higher prices in the market. Maybe, he says, the government might subsidize farmers for switching to organic practices.

A poignancy burdens these hopes. In Yashpal's office at the

cooling warehouse, the magisterial visage of Tikam Singh peers down from a large portrait hanging on the wall behind his desk. For three generations, the Singhs have planted potatoes here, helping build a regional agricultural powerhouse. But Yashpal worries his family's reign may be coming to an end. He confides that neither of his two children, both teenagers, have shown much interest in continuing in the business. They see the difficulty of farming these days, with the stress over the changing weather and environmental threats. But he hopes that if his new way of growing potatoes is successful, they may change their minds. Otherwise, he says, "It ends with my generation."

◈

The generational evolution of family farming on the Indo-Gangetic Plain is clear to see on the drive west from Agra, leaving behind the state of Uttar Pradesh and approaching the desert-like scrubland of Rajasthan. From generation to generation, farm sizes have diminished as parents divvy up portions to children. For example, a grandfather divides eight acres to give each of four sons two acres, which is then divided again to give two grandsons one acre each. Along the highways, the farms look like a collection of squares on a quilt. Rather than solid stretches of one crop on one farm, the crops and colors change rapidly from plot to plot. There are the golds and browns of wheat, the smoldering black char of burning rice stubble, the greens from leaves of spinach and lettuce and the spiky shoots of carrots and onions, the yellows from marigolds and mustard seed plants. Some farms are ornately embroidered with potato mounds; others are hemmed with fruit trees. On most farms, furrows run along the edges, filled with water for irrigation, electric pumps chugging away as they pull up the water from deep beneath the surface. Farmers along the way say underground water tables are falling a couple of feet each year, sinking to depths of 100, 200, 250 feet.

India may be the world's most water-stressed agriculture country. It has 18 percent of the global population and only 4 percent of the world's water resources. On the drylands of Rajasthan, water is liquid gold, much more precious than in the northern plains where canals bring runoff from the Himalayas. Here, agriculture is even more closely tied to the rhythm of the annual monsoons. And when that rhythm goes haywire, disaster ensues.

That is the story a group of eighteen millet farmers in the village of Nasirbaas gathered to tell.

As the middle of June 2022 approached, they eagerly prepared for the traditional onset of the monsoons in this part of the country. The government had raised its minimum support price for millet, a nutrient-rich grain hardier than wheat or rice, more suitable for the drylands. The farmers took the signal and committed to cover their several-acre plots in millet, envisioning handsome profits. The rains came as expected in mid-June and the farmers proceeded to plant. The monsoon would normally bring steady rains well into August, but after several days that June, the rain trickled to a stop.

Then nothing for many days, a string of forty-five to fifty in all. It was an epic drought in the middle of what should have been the rainy season. The millet, though sturdy, grew severely parched as the sun bore down. Some patches were clearly stunted, while others outright wilted and died. When the rains returned, haltingly, it was too little too late. In late August, the farmers began harvesting, to salvage what they could. They laid the stalks, moist from the late sprinkles, in their fields to dry.

And then the heavens opened with a tremendous deluge.

It rained continuously, without letup, for three days and three nights. For seventy-two hours, the farmers peered from their windows and leaned out of their doors to check the skies for any sign of a break. When they looked into their fields, they saw their harvested millet floating away. And the millet that was still standing, not yet harvested, was drowning as the water level accumulated on the sandy loam soil. That too was a disaster, for millet roots suffocate and rot in standing water.

For the farmers, it was a scene from the apocalypse. "I couldn't believe it. I had never seen anything like this before," said Ash Mohamad, the elder in the group. "We expect rains, light rains, but this was a downpour, pounding rain."

In those seventy-two hours, more than a foot of rain fell. That is roughly the average *annual* monsoon rainfall for the area, which would usually be spread over three to four months.

The crop was almost completely destroyed. They had expected a yield of more than a ton per acre, but they estimated they were able to sell at best 10 percent of what they could salvage. And what they brought to the market, stunted and frail, fetched just about two-thirds of the minimum support price. Making matters worse, the damaged millet they didn't bring to market was riddled with mold and couldn't even be used as animal feed. Their hoped-for windfall harvest became a wipeout.

The very day these farmers gathered to describe the extent of their losses to me, world leaders meeting in Egypt for the annual United Nations climate change summit agreed to provide "loss and damage" funding for vulnerable countries hit hard by climate disasters. Particularly extreme weather events that destroyed harvests and severely impacted farming communities would be a top priority. What befell the millet farmers of Nasirbaas just two months earlier seemed a model first case.

But no one here would be waiting for a check in the mail

from the United Nations. Instead, they turned to a higher authority.

"It depends on God," one of the farmers said. Many murmured in agreement.

One of the younger farmers, Kamal Kahn, noted that much also depends on people's actions. The climate was changing, he said, but too many of their actions were fixed, as if set in stone. The timing of their planting and harvesting, for instance. Their rigid crop rotations. Their assaults on the environment. "There is deforestation," he said. "Pollution is increasing from industry, all the factories, the stubble burning. We don't grow rice here, but still we get the impact. The ozone is thinning, the sun is hotter, temperatures are rising."

Surely, he said, there was something they, the farmers, could do as well. Ash Mohamad, the elder, counseled the group to look forward. "What has happened has happened," he said. "This is from God. It can either be a curse or benefit." Yes, benefits, he insisted. All the rainwater will lower the salinity of the soil, he said. It will increase the water level for the next crop. "We can take it as a positive."

NP Singh, an agricultural advisor of the SM Sehgal Foundation working with smallholder farmers on adapting to changing weather and environmental conditions, applauded this attitude. "You have to be positive," he told them. "As farmers you are dependent on nature." But even nature, he says, needs an assist.

Particularly in Rajasthan, where every drop of water counts. Here, the Sehgal Foundation, founded by Indian-born American crop scientist and philanthropist Surinder (Suri) Sehgal and his wife, Edda, is introducing new ways to conserve water and reduce rural poverty. Several miles down the road from Nasirbaas, farmer Samaydeen is leading his family into precision irrigation. He works three acres with his father Deenu. Deenu's father had twelve acres of land, which he divided by four for each of his sons. Samaydeen is harnessing new technology to make the most of their small farm.

Electricity is sporadic at best in the rural areas of Rajasthan. So Samaydeen has installed six sets of five solar panels, which provide the power to run a water pump. The panels, gleaming in the sun, are an unexpected sight on the scorching scrubland. Samaydeen knows that the same sun that brings oppressive heat can also bring relief. The panels are connected to a pump that brings up water from two hundred feet underground. The water flows through a network of thin pipes to provide drip irrigation to almost every inch of the family land that isn't covered by dwellings.

The water irrigates the cauliflower and cabbage growing under the solar panels. It irrigates his thirsty patch of wheat for one last season before he shifts to all vegetables. It irrigates his trial plots of onions, garlic, coriander, and fenugreek. Before harnessing the sun, he had to purchase water from a farmer with a well and a generator nearly two miles away. During the peak planting season, when the water was most needed, he was at the mercy of that farmer, who often claimed he had no more to sell because he needed it for his own crops. Now Samaydeen has his own reliable source of water; he's spared himself the expense of purchasing it, and even sells some of his water to neighboring farmers. He will no longer need to pay for the unreliable electricity. And the high-value vegetables will reap better prices than his wheat-millet rotation. All that, he calculates, will pay off his solar investment in a couple of years. On top of that are the savings to the environment: reducing reliance on carbon-based energy and conserving water by forsaking the traditional flood irrigation for the slower, and more effective, drip method.

"Water is my life," Samaydeen declares. "It has saved our farm."

In the nearby village of Baramada, farms have been created following the construction of a small dam and earthen pond by local village development committees with Sehgal Foundation support. Whether the monsoons are timely or not, the rain always becomes a benefit as it falls over a two-square-mile catchment area and flows into the pond behind the dam. Here, deluges are welcome. Local farmers quickly come and take the accumulated water to store for later use on their fields. Within a few weeks, any remaining water soaks into the sandy soil and the pond is dry until the next downpour.

The development committees report that the dam water has been used by six neighboring villages to bring more than one hundred acres of degraded land back to cultivation. Millet, wheat, and mustard now grow where once there was nothing but dust. In two years, the underground water table increased by about thirty feet, making wells more accessible. Residents who had been moving away from the area, particularly the youth, are now staying to work the land.

Gopal Kumar is one. He used to spend several months of the year in Punjab, picking cotton and working the rice paddies. Now, he has his own crops of cotton, wheat, and mustard in Baramada. "I'm able to stay home and feed my family," he says. He shakes his head in wonder. "This used to be barren land."

One village started an integrated guava and pomegranate orchard, planting onions, garlic, and eggplant beneath the trees. Having more fruits and vegetables for their children was the main goal, but the trees have also restored some balance to the environment. "We cut down so many trees," says farmer Deen Mohamad, "and then we complain about the changing climate. We keep making the same mistakes. If we cut down a tree that is twenty to twenty-five years old, even if we replace it with ten trees, it won't make up for the loss of that one tree for many years."

Another farmer, Ayyuab, also pleads for kindness to nature. He has set up a sprinkler system to irrigate his wheat and mustard. It gently sprays water over his crops from above. For one acre, he uses fifty-four sprinkler heads set up on poles, with each head calibrated to judiciously measure the water use. He says he now irrigates an acre in two hours; before the sprinkler, it would take up to ten hours. After watering one acre, he moves the sprinklers to the next acre. He says he uses only half the water he used in the old flood irrigation method, which indiscriminately spread water across his field whether it was needed or not.

As a child, he remembers, he could fetch water from the well with an outstretched arm. Now that same well, he says, has been dry for many years as the water level has steadily retreated. "We haven't been replenishing the water, just taking it. We need to take better care," he tells a group of farmers who have gathered to observe his sprinklers. "If we disturb nature, nature will disturb us."

Ayyuab recites a parable about a little bird with an upturned beak. Because of the shape of his beak, this bird couldn't drink water from the ground. He could only catch rainwater falling straight from the sky. One day, during a drought, the bird was really thirsty. He begged his father if he could go down to the river to drink. The father said it wouldn't work, for his beak turned up not down. But the little bird flew to the river anyway. There, try as he might, he couldn't position himself to drink. Frustrated, he returned to his father and complained about still being thirsty. "Why can't I drink from the ground?" "Well," said his father, "It's not in your nature."

Ayyuab laughs. The moral of the story, he says, is this: you can't go against nature. It will backfire if you try. Take what nature gives you and work with it, not against it. All farmers, he says, should take heed. Bending agriculture to your will, using modern forces to impose yourself, eventually doesn't work. Think of the deluge that went on for three days and destroyed the crops. Think of the rising temperatures that bake the fields. Climate change is nature's answer. The little bird couldn't bend his beak to defy what nature gave him. Neither should agriculture bend nature to its will.

He has learned, Ayyuab says. He is planning a shift from wheat and millet to vegetables, inspired by the experimenting of Kamal Kahn, the farmer who warned of the thinning ozone. Kamal has taken one acre of his land out of the grain rotation and has started to grow tomatoes. About 3,500 plants. The first year, he made rookie vegetable-grower mistakes. He didn't know that tomato plants could grow vertically instead of horizontally. He lost half his crop as the vines spread out on the soil and the tomatoes rotted while lying on the wet ground. Still, he made as much from the tomatoes as he would have with his grain rotation. The second year he installed trellises and the vines grew taller than him. He used organic mulch as fertilizer and as a check on weeds, and he deployed drip irrigation to control the soil moisture. He strategically placed yellow sticky pads and set pheromone traps among the plants to corral the pests. His profit doubled. Harvesting tomatoes was easier than cutting, drying, hauling, and milling the grain. His entire family joined in the work, saving additional labor costs. For his third year, he said, he would expand to two acres of tomatoes with about eight thousand plants, and diversify his vegetable patch to include radishes, eggplant, okra, and pumpkins. He has watched other farmers experiment with carrots. Maybe he would try them too.

Kamal is like the bird in his friend Ayyuab's parable, learning to be content with what nature gives him. "I'm earning more, I'm saving water, I'm not using chemicals, I'm feeding the family better," Kamal says. He is nourishing and preserving. "I'm very happy with my tomatoes."

As was another young tomato-growing farmer twelve time zones away, on the opposite side of the world, who was also questioning agricultural practices of the past.

CHAPTER 5
PAN-AMERICAN HIGHLANDS

Don't You Want To Know?

In Guatemala's western highlands, most everyone with a sliver of vacant land grows some maize. Maize grows in backyards and front yards, in schoolyards and churchyards, in courtyards, junkyards, brickyards, and up to the fences of graveyards. Maize was the most important crop for the Maya civilization and remains so for the descendants today; it makes up about three-quarters of their diet. Maize is usually served in some form for breakfast, lunch, and dinner, and the most ubiquitous snacks are corn chips—in lime, barbecue, chili, and picante flavors.

"Growing maize is what we do," says bus driver and maize farmer Rafael Perez as he stoops under the weight of a one-

Left: *Nourishing and Preserving*
Jorge Xivir and his tomatoes on the Guatemala highlands

hundred-pound sack of freshly harvested cobs. He gasps and then sighs in relief as he drops the bundle under his kitchen window. It adds to the work of a half-dozen family members who empty the sacks and rip the husks from the cobs, readying them for drying in the sun. Inside, maize tortillas are ready for dinner.

And then there is Jorge Xivir. Jorge is a full-time farmer about ten miles down the road and three hundred meters lower in elevation. His forefathers grew maize, of course. But he doesn't. Jorge grows tomatoes, thousands of them, on his small plots of land in the Zunil valley surrounded by a ring of volcanoes, including one called *La Muela*—The Molar—for its distinctive tooth shape. "Growing tomatoes is my passion," he says, sounding like Kamal in India. "I love seeing people eat them and smile." He grows tomatoes all year round. Maize, he notes, is only one season, "and that one is becoming more difficult."

While the ancient Maya communities revered maize as their evolutionary ancestor, today's Indigenous communities worry more and more about the crop's ability to survive. Growing maize that stands naked to the elements, watered only by rain for a single harvest, is becoming an ever-greater risk in these days of extreme weather events. Rafael says that in recent years his maize, grown by hand on a half-acre family plot in the foothills of the mountains, has been battered by hail, heat, drought, flooding, and wildly variable prices. Initial rains at planting time in April were followed by long periods of drought; then, at harvest time in November, he prayed for the rain to stop. Rafael wanted to break from the traditional reliance on maize, but was afraid that his highland soils could grow nothing else. And what, he wondered, would his family eat if he didn't grow maize? The diminishing returns from the staple crop were discouraging new investments in agriculture.

Rafael had watched some neighbors give up on their fields and seek other jobs and sources of income so they could buy maize on the market instead of growing it themselves. In the ultimate sign of spreading despair, the communities around Quetzaltenango, Guatemala's second-largest city and the capital of the western highlands, have been slowly depleted by migration northward in search of a better life. It has been the region of Guatemala with the highest levels of migration to the United States.

But other farmers like Jorge, in the lower valleys, have doubled down on agriculture, trying to outmaneuver the changing climate by switching crops and embracing new practices. In the past decade, they have turned the Zunil and Almolonga valleys into one of the most significant sources of fruits and vegetables in Central America. The crops that sprout from the fertile volcanic soil, naturally enriched with nutrients from lava and ash, are eye-popping specimens: radishes bigger than tennis balls, peppers larger than softballs, cabbages greater than volleyballs. And Jorge's tomatoes are bright red, robust, succulent—"the best-tasting ever." That's what his customers tell him.

Jorge's tomatoes grow in greenhouses; he began with one and within a half-dozen years had expanded to five. Those shimmering hoop structures stand like alien pods on these ancient fields, high-tech sentinels from the future. And they bear a message for the coming times: care for the environment, care for the trees on the mountain slopes, care for the natural springs that emerge from deep within those forests. For the water that flows from those springs, enabled by the forests, "is our life," Jorge reminds his fellow farmers. It was the harnessing of the mountain springs for agriculture that made the shift from maize possible. It allowed Jorge's grandfather to begin growing onions, and for his father to add carrots, radishes, turnips, and lettuce. And for him to specialize in tomatoes. Jorge, in his early thirties, sees the wave of migration, but he vows to stay put. "My family has been dedicated to agriculture for generations," he says. And he wants to make sure his isn't the last. "We must assure that we have water for the future. We can't assume it will last forever."

The water arrives at the farms from a network of flexible pipes that swing overhead from the mountain slopes to the valley floor. Every day, before he enters his greenhouses, Jorge looks up to the volcanic forests and wonders how long the trees will remain. Like the farms on the Indo-Gangetic Plain, the size of family-owned farms in Guatemala's western highlands has diminished through generations of hereditary division. They are now mainly a half-acre in size, one or two acres at most. They lie next to one another like dominoes, tightly placed end-to-end, covering the entire valley. Expansion isn't possible, nor really is further division. Jorge fears that the forests make a tempting target for farmers desperate to increase production. He has followed with dismay the developments in other Central and South American countries where agriculture expansion has laid waste to many forests, particularly in the vast Amazon River basin, and in so doing has dramatically altered the regional and global climate. And he knows it is happening in his country as well. According to Global Forest Watch, Guatemala lost more than four million acres of tree cover, a 20 percent decrease in total forest area, from 2000 to 2022. Jorge shudders at the thought that the forests of the Zunil and Almolonga volcanoes, and the springs they harbor, might vanish as well.

"We need to keep the trees and forests growing and thriving to conserve and protect the water, so the springs still flow," Jorge says. He rallies his fellow farmers every May to join in the municipal tree-planting day. It can be a hard sell. He estimates that maybe 5 percent of his neighbors are aware of the delicate balance between trees and the climate that preserves the water. "The rest think the water is always coming, that it will always be there," he says. "They don't have the information. I ask them, 'Don't you want to know?' They need to be made aware of what can happen if they aren't thinking ahead."

He offers his greenhouses as open classrooms, full of examples of how to nourish the community while preserving the environment. Made of sturdy plastic stretched over a wood and steel frame, Jorge's greenhouses form a protective shield against the assault of weather extremes and environmental threats like pests and disease. Each greenhouse occupies less than a tenth of an acre, a model of how to increase production without using more land. The rest of the space on his half-acre plot is left empty. Neighbors who fill most every inch of their open-air gardens with onions, carrots, cabbages, and chrysanthemums ask why he doesn't grow something there. At least plant some maize, they say. That would be the instinct of most Guatemalans. It's what his grandfather would have done. But Jorge, from a newer generation facing farming challenges unimagined by their elders, lets the land lie unused, as a *cordon sanitaire* around his greenhouse to keep any pests that might be attracted by maize away from his tomatoes.

Jorge cautiously opens the door to the first greenhouse he constructed, leaving just enough space to quickly slide through without letting in too many elements from the outside world. Here, within the bubble, Jorge is better able to control the changing climate and environmental challenges. Gauges regulate the temperature and humidity. Bags of compost, mulched biomass infused with soil-boosting microbes, cover the ground, enabling a more judicious use of fertilizer. Seven hundred or so tomato plants sink their roots in the compost and their vines poke through the plastic and twist their way up to the ceiling, wrapping around galvanized steel wires. He has designed a crop rotation strategy—two seasons of tomatoes followed by one of cucumbers—to regenerate the soil. Bees in box hives pollinate the plants; a blowing fan helps to spread the pollen. Jorge believes it is the bees that give his tomatoes a sweeter flavor favored by the market.

He kneels on the ground to show off the most important feature: drip irrigation to conserve water. A maze of thin plastic tubes releases water, infused with plant nutrients, into the compost on a controlled schedule: three times a week for twenty minutes each. Jorge calculates that the regulated dripping uses about five times less water than a sprinkler system, which is more commonly used in the valley. And sprinkling from above, he says, exposes the wet leaves to fungus attacks, which requires more pesticide use. To make the drip irrigation more efficient, Jorge uses plastic clips to prevent the vines from drooping away from the wires; bent vines, he says, restrict the uptake of water from the ground. He learned about the vine clips from watching YouTube videos.

Jorge inspects his tomatoes, which are just beginning to ripen. As he parts one thicket of vines, he comes face-to-face with a smiling picture of himself. It is at the center of a canvas banner bearing the logos of the U.S. Agency for International Development (USAID) and its Feed the Future Program. "Hope for the Future. A New Pathway for Farmers," the banner announces. USAID had funded a number of programs to help farmers transition to growing fruits and vegetables. Jorge was at the vanguard of the Innovative Solutions for Agricultural Value Chains Project, a ten-year, $75 million program working with thirty-five thousand farmers in the western highlands. Implemented by Agropecuaria

Popoyán, a Guatemalan agribusiness, the program aimed to foster crop diversification in the face of climate challenges, reduce poverty and chronic malnutrition in the western highlands, and lessen the flow of people to the U.S. border.

For Guatemala, like India, is an agricultural paradox. Most of the jumbo-sized vegetables of the volcanic valleys did not end up on the plates of the farmers who grew them or the plates of their neighbors. They planted, nurtured, and harvested foods that weren't part of their maize- and bean-centric diets. Instead, the vegetables were mainly packaged and shipped off to the urban centers of Guatemala and Central America or exported to wealthier nations like the United States. As a result, the lush fields of the highlands were surrounded by the worst malnutrition in the western hemisphere. Nearly half of all children under five years of age in Guatemala are malnourished and stunted in some manner. Here in the western highlands, childhood malnutrition is upwards of 70 percent.

Jorge hoped his tomatoes would put a dent in those woeful numbers. He learned how to get the most out of his greenhouses. Before the Popoyán project, he was producing about eight thousand pounds of tomatoes annually from his one outdoor patch. Now, in his five greenhouses, he harvests about seventy-five thousand tomatoes each year. That's a market value approaching $20,000, double the per capita income of Guatemala and an astonishing sum for smallholder farmers. "Jorge shows others that farming can be profitable. That you can grow another crop to mitigate the risk of growing only one, maize," says Popoyán project director German Gonzalez Diaz. He tells Jorge he is a trailblazer, with about ten thousand farmers in the western highlands using drip irrigation, composting, and new technology in the shift from maize to vegetables and fruit.

"Producers like Jorge don't want to leave the country. They have income, they have land, they have their families here. They don't want to go anywhere."

Jorge staggers his production in his five greenhouses so he harvests tomatoes every week of the year to sell in the market. "You can now find tomatoes in every season in every corner of the country," German notes.

That, Jorge insists, is his finest achievement. "I'm proud all this helps reduce malnutrition," he says, standing inside his first greenhouse. He invites customers to visit his greenhouses where he teaches them how to best feature the tomatoes in various dishes, and he explains how diversifying diets away from the reliance on maize and beans will help reduce childhood malnutrition and stunting.

"Here," he says, "they can see a vision of the future."

It is a future of forests and water and flourishing vegetables. It is a future of nourishing, preserving, and making a profit at the same time. A rare trifecta in the Pan-American highlands.

◈

In Latin America, we see both the potential and peril of our future. The region's highlands and grasslands, with their vegetables and fruits, their grains, potatoes, and beans, and their grazing cattle, have become indispensable to nourishing our population. And their mountains, rainforests, rivers, and coastal reefs, with a biodiversity nearly unrivaled in the world, remain indispensable to securing the planet's environmental health. Is a clash of these two indispensable forces inevitable, or is coexistence possible?

A seminal report released in 2014, *The Next Global Breadbasket: How Latin America Can Feed the World—A Call to Action for Addressing Challenges and Developing Solutions*, by lead authors Ginya Truitt Nakata of the Inter-American Development Bank and Margaret Zeigler of the Global Harvest Initiative, cast the region as key to both nourishing and preserving. It noted that in the second decade of the twenty-first century, the region—known in international economic parlance as Latin America and the Caribbean, or LAC—became the world's largest net food exporter. It produced 60 percent of the world's soybean exports, 45 percent of coffee and sugar, 44 percent of beef, 42 percent of poultry, 33 percent of maize, 70 percent of bananas, 12 percent of citrus, 13 percent of cacao, and varying percentages of a whole host of other foods that appear daily on tables all around the world.

And the potential to grow even more food for local and global consumption was enormous. The report noted that the LAC region is rich in three of the most important ingredients for agricultural production: land, water, and natural habitat. "The LAC," it continued, "has a third of the world's fresh-water resources, the most of any developing region when measured on a per capita basis. It has about 28 percent of the world's land that has been identified as having medium to high potential for sustainable expansion of cultivated area, and a 36 percent share of land that is within six hours travel time to a market. In fact, the region has more potentially suitable land for rainfed cultivation than the combined land from all other regions of the world outside of sub-Saharan Africa. With nearly a third of both the world's arable land and fresh water, Latin America may well hold the key to a solution to the world's food security challenges. Within LAC there is also enormous biodiversity, particularly in tropical areas, with the potential for producers in the region to make contributions to global advancements in medicine and agricultural science." In other words, who knows what great future discoveries that could benefit all of us—medicinal cures, superfoods!—are hidden away in those forests, marshes, reefs, and dense savannas. The potential value is incalculable.

But what will be the cost to all of us for unlocking this seemingly limitless potential? For the LAC is also one of the most biodiverse regions of the planet, with its vast forests, towering mountains, native grasslands, and rich coastal waters. It is home to about half of the world's land and water species. The Amazon rainforest, with its wonderous oxygen-producing abilities, is famously referred to as the "lungs of the world." The majestic Andes Mountains, with impressive glaciers and high-elevation soils, regulate the region's climate and have given the world foods like potatoes and quinoa.

Much of this has already been put in jeopardy during the LAC's agricultural ascendancy. To achieve the increased food production, farmers and ranchers expanded into natural habitats, converting them into agricultural land, decreasing the soils' water retention capacity, increasing carbon emissions, and shrinking the living space for the area's wildlife. According to *The Next Global Breadbasket* report, the LAC region represented about one-third of the global increase in land put into agricultural use in the five decades from 1960 to 2010. Over the same time, the region had lost an estimated 40 percent of its original forests, mainly to the advance of commercial agriculture—growing crops, grazing cattle—that enabled it to become a leading global exporter of soybeans and beef. Latin America's deforestation rate soared to triple that of the rest of the world; between 1990 and 2015 it experienced the biggest loss of forest area of any region, some ninety-seven million hectares (an area larger than Texas and Oklahoma combined). At one point, Paraguay's El Chaco was the fastest disappearing forest on the planet, losing about three million hectares in less than ten years. The World Wildlife Fund, in its 2022 *Living Planet*

Report, estimated a 94 percent biodiversity loss in Latin America since 1970. It reported there were declines in all species, but most significantly in freshwater fish, reptiles, and amphibians.

A growing awareness of the environmental cost, along with government and private sector pledges to reduce carbon emissions and better monitoring of deforestation practices and patterns, slowed the pace of deforestation in subsequent years. But that didn't hold, as new governments in some countries retreated from their commitments and encouraged opening new land in the pursuit of economic expansion, hoping to cash in on the global demand for grains and protein. In 2022, the Amazon rainforest faced its largest deforestation in almost two decades, according to a deforestation survey by the World Resources Institute. Brazil had the highest rate of deforestation globally, accounting for more than 40 percent of the world's tree loss. In Bolivia, forest loss jumped by 32 percent that year.

All of this has made the LAC one of the leading contributors to greenhouse gases driving changes to our climate. Over a fifteen-year period at the start of this century, the destruction of Latin America's forests was the world's fifth-highest source of global carbon emissions. And the large concentration of cattle made the region one of the leading emitters of methane.

The impacts are striking close to home. The countries of Central America, occupying a narrow strip of land sandwiched between the Atlantic and Pacific oceans, are considered to be among the most vulnerable in the world to weather extremes such as hurricanes and flooding. The warming of those nearby surface water temperatures influences the El Niño and La Niña weather patterns that spread across the world. Temperatures in the Andean region are rising faster than in most other locations;

warming mountain conditions are causing glaciers to melt and forcing farmers of high-elevation crops like potatoes and coffee to alter growing seasons and move to even higher ground.

As in Africa and India, the family farmers and smallholders of Latin America are absorbing the greatest impact of climatic changes and are also leading the way in deploying innovative solutions to both increase food production and preserve the environment. Family farming accounts for 80 percent of all farms and 40 percent of all food production in the LAC region. They have made Latin America the global leader in no-till farming, a cropping method that minimizes soil disturbance in planting and harvesting. Crop residue is usually left in the field, adding nutrients to the soil, holding water, and decreasing erosion. The ranchers in the region have been pioneers in developing silvo-pastoral systems that integrate trees, forage, and livestock to both increase beef and milk production while reducing the impact on the environment. It is a potentially carbon-neutral alternative to cutting down forests to create more grazing space. Those ranchers report a reduction in the need for chemical fertilizers (they rely on cattle dung instead), more productive soils, and the ability to graze more animals on the same amount of land. Consumers consider the dairy products to be tastier and more nutritious. Environmentalists have recorded increased biodiversity of insects, birds, and plants, and increased carbon intake by the soils.

The spread of these practices will increase in importance in coming decades, given projections that nearly 50 percent of the world's new cropland by 2050 will be in Latin America. It is this calculation that has both the multinational food industry and environmentalists closely watching the vast tropical savanna of eastern Colombia known as Orinoquia.

This second-largest stretch of tropical grassland in South

America is largely virgin agriculture territory, virtually untouched by the plow, having largely remained beyond development during five decades of civil conflict that raged in the countryside. Land use in the region has been primarily small-scale cattle ranching and low-input traditional farming. But with the signing of a peace accord in 2016, these pristine grasslands that roll to the Amazon rainforest have been identified as a priority area for post-conflict development by the public and private sectors. The Colombian government considers Orinoquia the last great frontier for economic growth. The national poverty rate in Colombia was near 40 percent in 2022, with 12 percent of children under five stunted from malnutrition. Orinoquia is seen not only as a potentially bountiful source of food to meet local and global demand, but also a source of jobs and income. Initial ambitions focused on converting up to fourteen million hectares—more than half of the region—for large-scale agriculture use, focusing on soybean and palm oil production as well as cattle. As one local government plan stated: "We can be the agricultural pantry of this country and be much more competitive. Raising per capita income, lowering the poverty rate, and solving unsatisfied basic needs are urgent tasks."

Orinoquia is also one of the world's treasures of biodiversity, valued by environmentalists for the same untouched qualities so attractive to the agriculture sector. Colombia, by some estimates, hosts 10 percent of the planet's biodiversity, and is particularly prized for its plant, bird, butterfly, and amphibian species. About half of the country is covered by forests. Orinoquia is the center of much of this biodiversity. According to the World Wildlife Fund, more than one-third of its species are unique to the region. It produces 30 percent of Colombia's fresh water. It is the location of nearly 50 percent of the country's continental wetlands where cherished species such as the tapir, jaguar, and caiman live. And it is home to more than twenty Indigenous ethnic groups and a ten-thousand-year history of settlement.

Environmentalists have warned that large-scale development could create unintended consequences for the very soil, biodiversity, and water upon which agriculture production wholly depends. They argue that land conversion calculations need to account for the critical role Orinoquia plays in sequestering carbon in the virgin soils and in regulating the temperature, rainfall patterns, and soil regeneration of the region and the country.

Already agriculture development since the peace accords has shown Orinoquia's food production potential; it supplies about 30 percent of the country's crop production and 20 percent of the cattle herd. However, the World Bank calculated that because of agricultural activity, the natural ecosystems of the Orinoquia region were disappearing at "an alarming rate."

In an effort to avoid mistakes of the past playing out again in Orinoquia, the World Wildlife Fund and the World Bank joined with Colombia's National Natural Parks and the Ministry of Environment and Sustainable Development to launch the Orinoquia Integrated Sustainable Landscapes project at the end of 2021. Its goal is to ensure the conservation of the region while promoting its development—a high-wire balancing act that has rarely succeeded anywhere else. The stakes are high and the questions abundant. Are the lessons of other ancient lands being heeded? Can actions be taken that are good for both the population and the planet? Can it be a template for creating a brave new world where agriculture and conservation work together as allies? Like the Kapiti Ranch in Kenya, Orinoquia may be a proving ground for whether nourishing and preserving can coexist.

◈

A sudden autumn hailstorm, striking out of the blue, pelted the potato patch high in the Peruvian Andes. A strong wind followed. And then bitter cold and frost. Farmer Gladis Dina Rurush Jorge wondered what was happening. It was May in the southern hemisphere, the dry season heading into winter. It shouldn't be so wet or cold, Gladis thought. And what about the hail, a rarity at any time of the year? She hurried out to her potatoes and found half of them flattened from the storm. The plants had just begun to flower, providing nourishment to the potatoes developing in the ground. Gladis snapped photos on her cell phone, recorded a video, and quickly dispatched the evidence of the damage to a WhatsApp group.

"There's the potatoes. They are blooming," she narrates in the video. Her voice is soft, incredulous. She repeats her descriptions, as if trying to comprehend what she is seeing, while the camera surveys the scene. "But yesterday there was a strong hailstorm and it has mistreated [the field] in this way. It has totally destroyed it, like this. Some of them. Some of them are still standing."

The camera focuses on the damage. "Here, always the hail. And the wind, very strong. Some of them have been saved, but some of them not. Look how it is flattened. I don't know if it will be able to recover. It was just blooming. But it has been mistreated by the hailstorm. Everything, all the way down the hill."

The camera finds some plants still standing. "Thank God this part is beautiful."

Then she looks up the hill, seeing the offending balls of ice on the ground. Still incredulous. "There is still hail in the upper elevations."

Immediately, another farmer wrote back: "Gladis, don't be sad. We can help each other. It would be good for us Guardians who are having problems with the climate or disease to share so we know who needs help. Greetings from the Guardians, and we continue forward."

A second farmer replied, "Oh, what a shame, colleague. Be alert, colleagues, of the hail."

A third said: "Fellow Guardians, greetings to everyone. Yes, the climate change is very strong. Have strength, my colleagues, we will help each other to document the climate change."

The Guardians are members of the Association of Guardians of the Native Potato of Central Peru, a group of Indigenous farmers from dozens of highland communities who have given themselves the task of preserving potato diversity. They represent about fifty families of small-scale farmers who have been growing potatoes for generations, protecting them as a cherished inheritance stretching back centuries. They live at very high elevations, about 3,500 meters above sea level, in communities where the world's potatoes originated. Each family plants fifty to three hundred different varieties and acts as custodian of the evolutionary legacy of exceptionally rich collections of native potato varieties.

Their name is evocative of the Guardians of the Galaxy, the superheroes of Marvel Comics. But these farmers are truly real-life superheroes, fighting climate change and saving the origins of the humble, hardy potato from extinction. For if these native varieties were to ever vanish, so would precious genetic material of one of the most-consumed foods in all the world. Throughout history, since the first tubers were taken from South America to Europe in the sixteenth century, potatoes have saved millions of people from hunger. They have often been the bulwark against famine, the last crop available when others are wiped out by drought, flooding, or excessive heat or cold. But now, it is the potatoes struggling to survive. They are vulnerable to weather extremes like those Gladis experienced, and to disease, as the

Irish experienced in the Great Famine and as Yashpal Singh is experiencing now near Agra, India.

The Andes are home to at least 4,500 types of native tubers. With relatively high levels of zinc, iron, potassium, vitamin C, and antioxidants, they are fundamental to the health of Indigenous communities. And now, potatoes are the canaries in the coal mine of climate change. The Guardians fear that if one of their native species goes, many other species will follow. That would be an incalculable loss to all of us. Ginya Truitt Nakata, who moved on from the Inter-American Development Bank to the International Potato Center based in Lima, Peru, noted that the Andes' potato diversity is used by scientists to breed nutritious, disease-resistant potato varieties around the world. "Just one of these varieties," she added, "could hold the key to alleviating the next global hunger crisis."

"This is our millennial wealth," says Martha Mamani Condori, another Guardian. Her "our" means all of us. "People should know that we have a millennial richness here, not only in our potatoes but in our culture and customs as well."

One month after sounding the climate alarm on WhatsApp, Gladis shared more experiences on a Zoom call with Martha and another Guardian, Victoriano Fernandez Morales. They represented three outposts on the potato crescent of south-central Peru: Gladis in the Ancash region, Martha from Cuzco, and Victoriano from Huanuco.

"I have been really impacted this year," Gladis reports. "Two of my plots were destroyed by hail and cold. Fortunately, I planted four plots. So I lost 50 percent." A delay in the spring rains had set back the planting; without that delay, the potatoes would have been farther along when the unusual hailstorm hit, and the damage would have been less. The later harvest meant the potatoes that survived were late to the market. "In past years, I would have sold everything by now," she says. "So as you can see, climate change has affected everything."

Victoriano adds that it was hard to see anything clearly these days. "We can't see the different seasons of the year anymore. Before we had winter, summer, spring, autumn; they had certain dates. Now, no matter the day, it can rain. Sometimes too much. Other days, too much heat. That's not good. We have drought." There had been a rhythm to the year: the rainy season was December to March, the dry season May to September. Traditionally, he would plant in July and August. But that was now stretching into September and October. "This year," he says, "I even planted in November."

He continues: "We're changing the calendar of crop seasons, because we can lose crops in some months. As Guardians, we need to continue conserving and seeing the realities that we're living in. We're seeing that in this era of our times, we need to adjust our calendar."

"We're trying to see if we can plant in different rhythms, to save our seeds," Gladis adds. She notes that a lost season means more than lost food. It also means lost seeds to plant the next year.

To compensate for weather losses, they are also planting in wider areas, with more seeds. And, Victoriano adds, "I'm starting to grow higher up the mountain." For years, he and the other Guardians would grow 3,000 to 3,500 meters above sea level. Then as the seasons shifted, and the temperatures increased, they moved up to 3,800 to 4,000 meters. They also tried to outrun pests, like weevils, that proliferate in warmer weather.

The potato farmers of Peru aren't alone in taking radical steps to counter changing weather patterns. In Honduras, some corn and soybean farmers have turned their rhythm upside-down. It became too dry at the traditional time to plant corn and too wet at the start of bean planting. After suffering damage over several seasons of each crop, they flipped seasons. And they constructed a communal rain harvesting network of retention ponds to preserve water from the wet season to help parched crops during the dry spells.

In Africa, the coffee farmers on Mt. Kenya and other peaks have also been climbing higher up the slopes in search of cooler temperatures and healthier soils. In Asia, farmers are abandoning traditional crops for newer ones. In India, they have switched from the Green Revolution's wheat and rice to fruit and vegetables, and the same is happening in Bangladesh. Farmers in Vietnam, where rice flourished in the Mekong Delta, are moving out of rice and instead raising shrimp in waters that have become too salty as rising sea levels displace fresh water in the Mekong River. The decades of intensive rice production in a number of places in the world pushed old varieties out of markets, reducing the great genetic diversity of rice. Now, researchers are going back to those ancient strains, at least those that have been preserved, to breed new varieties that can perhaps withstand the forces of climate change and thrive in heat, drought, excessive flooding, and even in salt water.

This practice, of using ancient crop varieties to produce new ones to help nourish the planet in the face of climate change, amplifies the importance of the Guardians' work in the Andes, along with the efforts of International Potato Center scientists, to preserve the native varieties of potatoes. They know it is a delicate balance. The higher up they grow and the more they expand their fields to stay ahead of the changing climate, the greater the toll on the environment. They have begun planting in virgin soils, undisturbed over time, that have sequestered carbon from the atmosphere for centuries. As they sow their potatoes in those soils, the carbon released adds to the buildup of greenhouse gases that lead to the warming that is forcing them further up the mountain in the first place.

"We are very conscious of taking care of the environment," says Victoriano. "We have to care for the land and water."

And the forests. Victoriano explains how he is nurturing his own forest on his land as he sees trees felled on the mountain slopes to accommodate population growth. There is increasing urbanization, and agriculture expansion to feed the additional people. This is shrinking the natural habitat of the wildlife and pushing animals into the potato patches in search of food. Like humans, they also depend on the potatoes in times of drought.

"They try to dig up the potatoes, eat the flowers and leaves," Martha says. It is happening more as rains fail and heat intensifies. "If there is a drought, the animals don't have anything to eat."

The Guardians are also battling a more insidious foe: a culture that takes their work for granted, and the impacts of climate change as well. "Sometimes we get discouraged because it's not valued to plant and the [native] potato isn't so commercial," Martha says. Some varieties, she notes, are disappearing because there is no market demand and farmers give up on growing them.

The traditional varieties nurtured by the Guardians are a rainbow of colors: red, yellow, purple, brown, black, mottled. And an array of shapes and sizes: big, little, round, elongated. With this diversity also comes greater nutritional value, as each native variety has its own nutrient and mineral composition. But it is also this diversity of color and shape that makes them less desirable in the market; consumers generally prefer their potatoes to be one color, one size, one shape.

The Guardians are aware of what has happened to another ancient crop of the Andes: quinoa. As the edible seed was embraced by consumers in the U.S. and other wealthier nations as a superfood prized for its rich range of minerals and amino acids, the market zeroed in on just a few uniform strains. Farmers took this market signal and focused on planting the more commercial varieties to meet demand, potentially narrowing quinoa's genetic diversity.

The potato Guardians warn that the world shouldn't let that happen to their native varieties. They would like support from the market. "If nobody buys from us, we have no income to continue," Gladis says. "If we don't continue to produce our potatoes, maybe the public won't have this rich food anymore." She adds wistfully, "We only live from this potato, that we grow and harvest."

Victoriano warns that some potato farmers are leaving their fields to find better income with other crops, or other professions. "There's a migration of people from rural to urban areas. Too many people leave their plots and fields," he says. "They just leave. We Guardians in the rural areas, our efforts must continue."

He sees the disaffection even in the Guardians' own families. "Our sons and daughters, there is no support for them. They emigrate to the cities, change professions. We need to make sure people don't leave the communities."

Martha, one of the youngest Guardians, in her early twenties, tries to rally her generation to join their mission to preserve the native potatoes. "Me as a young woman, I want to influence other young people not to lose hope, don't get discouraged. They say they don't want to continue the legacy. My father has been involved in this his whole life, he dedicated his life to preserving the potatoes. He inculcated this in his sons and daughters to continue. I want other young people to get involved and continue."

She believes the Guardians' efforts could be made more attractive to younger people if there was more investment in their farms from the state and local government and agriculture institutions, as was promised in the International Treaty on Plant Genetic Resources for Food and Agriculture, adopted in 2001. The treaty acknowledged the concept of "benefit sharing," in which the families and communities preserving the diversity of native varieties would be recognized and compensated for the work they are doing—a "fair and equitable sharing" of the benefits arising out of their use, in the words of the treaty. More than two decades later, little support has reached the mountain communities of the Andes. So the Guardians organized themselves to support each other.

Their task at the top of the Andes can feel like a lonely, isolating one. "People should value the efforts of planting potatoes and other tubers, our efforts in confronting climate change, beginning with preparation of the soil, planting, growing," Martha says. "It's a big effort of not just one person, but a whole family. It's an entire year. An entire profession, what we do. The only way is to go forward. It's our major livelihood. If we are valued, we'll feel encouraged to keep going in the face of climate change and difficulties."

Adds Victoriano, the veteran of the Guardians with seven decades of wisdom: "That will be the salvation for us, and the potatoes." And much more. "We know potatoes help to save the hungry of the world."

In the effort to nourish and preserve, they are certainly not alone. For a continent away to the north, on the American Great Plains, another battle to carry on a generational farming legacy in the face of a changing climate rages.

DeWALT

CHAPTER 6
THE GREAT PLAINS

Why Do We Still Do That?

As spring turned to summer in 2023, the fabled amber waves of grain on America's Great Plains were looking more like toasted mini-wheats. The hard red winter wheat, which produces the flour that turned the prairie into the world's breadbasket, was dying of thirst. In central Kansas, there hadn't been any precipitation to speak of for more than a year. Drought had been creeping across the state, county by county, since the fall of 2021. There were scattered downpours in the spring of 2022, but that was basically it for the spring and summer. The winter wheat seeds were sown, as usual, in the fall, in the hope that some rain would appear. But it remained elusive. The drought only deepened. The expected thick cover of snow during the

Left: *Where the Buffalo Roam*
TJ Heinert restoring the Lakota herd

winter months never arrived, so there was little moisture to water the seeds as the ground thawed in the spring. There were no April showers, and very few in May. The U.S. Drought Monitor reported that almost 60 percent of Kansas was experiencing "extreme" or "exceptional" drought conditions. The wheat sprouts that pushed up through the parched soil were stunted, the heads of grain on top of the stalks were meager.

By May, when the amber wave should have been in full force, Kansas farmers were abandoning the crop in their fields at a historic rate. Some were intentionally dousing their wheat with crop-killing chemicals to spare themselves the expense of harvesting it, reaping insurance payouts instead. Others were plowing it back into the ground to at least strengthen their soil for the next season. Still others were cutting the wheat early to mix with cattle forage. The U.S. Department of Agriculture, in a grim estimate, predicted Kansas farmers would abandon nearly 20 percent of their crop; nationally, it noted that total winter wheat abandonment could be as high as one-third of the planted crop. That would be the highest rate in nearly a century.

Throughout the growing season, Brandon Kaufman forlornly surveyed the fields of central Kansas. "The wheat should be twice as tall as it is," he notes as June arrives. He and neighboring farmers are bracing for a diminished harvest. "Maybe twenty-five or thirty bushels an acre," he predicts. "We used to get fifty bushels with our eyes closed."

Several generations of Kaufmans have grown wheat here. Now, looking upon the stunted crop, he sees the toll that year after year of plowing has taken on the land. He points to gullies running through a neighbor's field that he watched deepen each season as erosion carried away precious topsoil. The organic matter and nutrient structure of the soil—so important for holding water, nurturing the vital underground life of insects and micro-organisms, and sequestering carbon from the air—has deteriorated over time. The persistent drought seems to confirm the many warnings that something sinister is indeed happening to the climate and ecosystem. This isn't like the drought of the 1930s that arrived out of the blue and brought the cataclysmic Dust Bowl and years of hardship with it. The pitiable state of the Kansas wheat crop is the result of something more than a sudden weather extreme; this is an inexorable event that has been building for a long time in the atmosphere and in the soil underground. For Brandon, it is, literally and figuratively, the last straw.

"I'm going to get a divorce from wheat," he proclaims.

It is quite a decision for a young farmer whose forefathers wed this land to wheat 150 years before. His ancestors were among the immigrants from Germany and Russia who began arriving on the Plains of Kansas in the 1870s with hard red winter wheat seeds sown into the hems of their clothes and stuffed into their children's dolls for safekeeping on the long journey. It was those seeds that transformed the prairie from a natural wonderland of tall grass and bison into an agricultural behemoth of row crops and cattle. It was that wheat that made Kansas and the American Midwest a food powerhouse. It was also those wheat seeds that set the Great Collision in motion on the Great Plains, and positioned America as the leader of a global agriculture industry that has devoured land and blanketed it with chemicals to increase production, prioritizing feeding over preserving. It was the wheat that forever changed the character of the Plains, bent the prairie to the force of the plow, and led to the Dust Bowl that carried away millions of tons of precious prairie topsoil, setting the stage for a whirring cycle of booms and busts in the century that has followed.

And now, generations later, a descendant of those wheat-pioneering immigrants is becoming a pioneer of his own. Brandon isn't about to give up farming. Quite to the contrary. He is more passionate about curating his family's land, near the town of Moundridge, north of Wichita, than ever before. Instead, he is setting his farm on a different course, in essence turning back the clock to a time before the wheat arrived, to a time when providing for the population and preserving the land for the future was not only possible but preferred.

"I'm going to allow things to grow and plant into it and not terminate the growing crop until after I have establishment," he says. Rather than planting and harvesting an annual crop of wheat—assaulting the ground with discs and plows and the immense weight of machinery every year—he has begun to nurture a variety of perennials. Once planted, perennials keep growing year after year, yielding multiple harvests. They allow roots to burrow deep into the ground, restoring soils, holding them in place, rebuilding organic matter, and recapturing original elements of the prairie. His hope is that these perennials can be developed as food grains for human consumption, such as the ancient intermediate wheatgrass now known as Kernza, as well as fodder for grazing cattle. He will grow the perennials together in a polyculture of crops rather than in an isolated monoculture. Approaching a thirteen-acre patch of Kernza he planted a couple of years earlier, now a swath of green amid the brown of the drought, he ventures, "This is a good place to try something new."

His inspiration comes from the daily wonders of the natural world. "How did Mother Nature manage all these years?" he asks. "Mother Nature has tens of thousands of years of research and development. Why do we need to recreate what she already knows? We're trying to mimic Mother Nature."

On another central Kansas farm, a few counties over,

another descendant of the hard winter wheat pioneers pondered the same thing. "I would love to get back to the place of farming with nature as opposed to trying to destroy every remnant of the ecosystem there is," says Jason Schmidt, a dairy farmer. "I look toward the prairies to model how I graze and attempt to do some replicating of what was here on the prairie."

A few days earlier, he had cut his father's struggling wheat far ahead of the normal harvest time; insurance payments would recoup some expenses. Jason rolled the straw into bales, which rest beside his house to be added to cattle forage in the fall and winter. It won't end up in a loaf of bread.

"We're at least fifteen inches of rain behind for the year," he says. He walks to a picnic table on the other side of his house, under the shade of maple and mulberry trees near the barn. His phone pings with a series of urgent weather alerts. A thunderstorm with lightning has been spotted in the vicinity; radar indicates it is rapidly approaching Jason's farm, expected to hit within the half hour. Jason looks up at the clouds gathering in the distance and sits at the table. In past years, he might have scattered for cover, like Dorothy and her dog Toto scurrying for shelter in the face of a Kansas tornado that lifted them to the Land of Oz. Now, during this drought, he is calm. He'll believe it when he sees it.

His dog Pepper bounds enthusiastically behind him and curls around his feet at the picnic table. "My climate change dog," Jason says. Pepper is an Australian cattle dog, also called a Blue Heeler for its tendency to nip at the heels of reluctant cows. Pepper has a shorter, less furry coat than Jason's other breed, the Australian shepherd, which carries a denser undercoat. He notes that Pepper does better corralling cattle in the hotter weather.

Residents of the Great Plains have grown accustomed to the

prospect of sudden, violent summer storms. Jason has become keenly aware of the ominous, slow-building nature of climate change. Despite the pinging alerts, the half-hour storm warning period passes, as does the entire afternoon, with no rain on Jason's land. "It's become so erratic, isolated rather than the large rain events we need," he sighs, disappointed. "Some people get something, but a lot of people don't."

Like many of the farmers we have met on other ancient growing lands of the world, Brandon and Jason rue how modern agriculture's stewardship of the land has often gone awry. They wrestle with how their own, and their families', past stewardship, no matter how careful, has unleashed unintended consequences. "It's a weird thing to be proud of family history, of agriculture history, and have a sense of place, but to also know it's my ancestors who plowed up the prairie," Jason acknowledges.

The introduction of wheat brought great value to the American prairie, but it also precipitated its declining fortunes. For along with the annual wheat seeds came the relentless turning of the prairie soils, the monocropping, the erosion, the release of carbon and greenhouse gases from the fields and cows. Eventually, the land turned against them in the Dust Bowl, long considered the greatest human-caused ecological disaster in American history. But now, as the forces of climate change gather, Brandon and Jason worry: what even greater disasters loom ahead?

"I'm not unhappy with my forefathers for using the plow and steel, that's what they knew," Brandon says. But he now deals with the consequences every day. He notes that when the ancestors first settled on the prairie, their land was among the richest, most fertile in the world, with a 6 to 7 percent level of organic matter in the soil. Organic matter is the fraction of soil composed of material produced by living organisms—plants or animals—in various states of decomposition. It provides nutrients essential for plant growth, such as nitrogen, phosphorous, and sulfur. It nurtures diverse soil microbes that help fight pests and diseases and provides an easier path for roots to penetrate the soil to find water, air, and nutrients. Now, Brandon estimates, some of those soils are down to 1 to 2 percent organic matter.

"We're working with a degraded resource," he says.

Degraded. It was stunning to hear that word on the American Great Plains. I mentioned that when Abebe Moliso in Ethiopia talked about degraded land, he meant soils so barren and so scarred by erosion that absolutely nothing would grow, forcing his family to abandon their ancestral farming ways. Abebe would certainly marvel at the fertility of these fields, even with the wheat in its stunted condition. The Plains of Kansas have a long way to go to reach that stage.

Of course, Brandon acknowledges. But, he quickly adds, "We've only been here 150 years." Without change, he says, "We'll eventually end up like some of the lands of Africa."

And so Abebe's question echoes from the Great Rift Valley across the Great Plains: Why would we do that?

Like Abebe, Brandon and Jason were asking questions as they charted new futures for their lands. What really are we doing? Are we just going through the motions, carrying on the same practices year after year? Why do we farm the way we do? That's the way my father and grandfather farmed. Why do we still do that?

And like Abebe in Ethiopia, their farming in Kansas has become all about discovering, experimenting, listening. "The wisdom is in the tall grass prairies, and the fact that that ecosystem was a balanced ecosystem that survived in this climatically volatile area for eons," Jason says. "So what is the wisdom of that prairie? I think we can take any multitude of lessons from that and how we manage our lands."

"You have to go out and learn and educate yourself," Brandon insists. "We all have to try something different."

Before it's too late.

◈

Dramatic climate extremes through the centuries—from the Ice Age to the Dust Bowl—have chiseled the Great Plains into a defining feature of North America, a vast swath of flatland rolling down from the Canadian provinces of Alberta, Saskatchewan, and Manitoba into the Dakotas and Montana and then straight down the center of the continent to the Oklahoma panhandle and Texas. Much of the geography of the Great Plains is a result of sediment buildup and erosion since the Cretaceous period, when a vast, shallow body of water known as the Western Interior Seaway covered the area. Over time, rivers running down from the Rocky Mountains to the west created alluvial fans with additional deposits of soil, gravel, and sand. During the Pleistocene epoch, which began two million years ago, great continental ice sheets slid down from Canada. These Ice Age glaciers flattened and smoothed the land of the northern Plains, carved out lakes, and left additional layers of sediment deposits. When the last glaciers melted around ten thousand years ago, a magnificent prairie emerged. Tall grasses and brilliant flowers flourished and attracted a diverse array of animals and birds. For centuries, this wildlife thrived, relatively undisturbed.

Great herds of bison, elk, and deer dominated the landscape. Human population was sparse until about eight hundred years ago, when Native American tribes from the north, east, and southeast regions moved to the Great Plains in larger numbers to hunt bison for food, shelter, tools, and clothing. They were mainly hunters and gatherers, living in rhythm with nature. They gathered wild fruits, grains, and vegetables growing on the prairie. As some of the tribes settled, they began raising crops like corn, beans, and squash, reaping what the rich soils would give them. They were careful not to overhunt the bison and take more than they needed to survive. As a result, the bison population thrived and multiplied, with some estimates placing the total size of the American herd by the middle of the nineteenth century perhaps as high as sixty million.

And the ecosystem thrived as well, for as the bison roamed the Plains, they massaged the land, mixing their nutrient-rich manure into the soil as they moved. And their innate grazing nature curated the grasses with an instinctive pruning, nibbling not all the way to the ground but only as much to stimulate further regenerative growth. This allowed the prairie grasses and wild grains to blanket the land and anchor their roots deep into the ground, holding the soil in place against the winds and rains. So vast were the animal herds that the prairie often took on a dark hue and reverberated with a sound that became known as the Thunder of the Plains. As the explorers Lewis and Clark came upon the majestic herds while charting the western territories during their 1804–06 expedition, they could hardly believe their eyes. In their journals they noted, "immence herds of Buffaloe deer Elk and Antelopes which we saw in every direction feeding on the hills and plains."

That expedition fueled the nation's appetite to expand, and within several decades, traders and caravans of settlers were crossing the Great Plains headed to western outposts, with stops along the way to form new settlements. They traveled under the banner of Manifest Destiny, the nineteenth century belief that America had a sacred duty to settle the continent by westward expansion, imposing dominion over all the land and the Indigenous peoples along the way. While climate extremes shaped the geology of the Great Plains, this philosophy shaped the demography, along with two acts signed by President Abraham Lincoln in 1862, during the Civil War: the Pacific Railways Act, which envisioned linking the east and west coasts with trains steaming from the Atlantic to the Pacific, and the Homestead Act, which granted adult heads of families who had never taken up arms against the U.S. government 160 acres of surveyed public lands, so long as they lived on and cultivated the plot.

Both of these acts accelerated the settlement of the western territory. Within four decades, the Native American way of life and the great herds of bison were on the verge of extinction. To eliminate the Native Americans, the government set out to kill that which gave them life—the buffalo, or *Tatanka*, as the Lakota call them. The two U.S. Army generals who led the "scorched earth" strategy in the South to end the Civil War, Sherman and Sheridan, were put in command in the West, responsible for protecting the railroads and the new settlers. A ruthless era of "Indian wars" and "The Great Bison Slaughter" turned the peaceful prairie into a stage of vast carnage.

The railroads became an ally of the army, even advertising "hunting by rail" adventures. Huge buffalo hunting parties steamed into the Great Plains. Hundreds of hunters rode the Iron Horses, taking position on the roofs and pointing their rifles out the windows, picking off buffalo grazing alongside the tracks. The Native Americans who survived were shuttled off to reservations, usually situated on the most unproductive land. The few remaining buffalo were herded into a new national park, Yellowstone. By 1900, the tens of millions were down to only about three hundred.

As the Native Americans' way of life and their great herds of buffalo died, so did the character of the prairie as previously known. For the railroads also brought hordes of settlers to the Plains, enticing them with a siren set of advertisements promising adventure—and profit. Like this one from the Atchison, Topeka & Santa Fe Railroad:

"If you want a farm or home. 'The best thing in the West.' Atchison, Topeka & Santa Fe Railroad Lands in Southwest Kansas. A Start on the Prairie. The Same Place after Six Years Work and Profit. Temperate Climate, Excellent Health, Pure and Abundant Water. Good Soil for Wheat, Corn and Fruit. The Best Stock Country in the World."

New arrivals rushed to the prairie, bringing steel implements to bust the sod and overturn the soils that had been largely undisturbed through time. Their mission was to shape nature to their purpose—dominating and conquering it according to Manifest Destiny—rather than working with nature, as the Native Americans had.

Among the newcomers were Brandon's and Jason's ancestors, and their wheat seeds. They had been farming in southern Russia, primarily in the area of present-day Ukraine, for about a century, having accepted the invitation of Catherine the Great to western Europeans to develop the steppe into a rich agriculture zone. She welcomed Mennonites coming from areas of today's Germany and Holland with promises that they could govern their own communities and follow their pacifist beliefs. With their efforts, they turned their new home into a prolific grain belt.

But by the 1870s, as nationalism grew in central Europe and the Russian government reneged on the special arrangements with the German-speaking farmers, the Mennonites were on the move again. They sent a group of scouts to explore new lands, particularly in North America. One of those explorers, a young man named Bernhard Warkentin, son of a prominent grain miller, came to the United States in 1872. As he made his way through the Midwest, he considered the possibilities: Wisconsin, too many trees; Minnesota, too many stones and lakes; the Dakotas, wide-open windy spaces. Then he arrived in central Kansas. The landscape reminded him of home; he saw a place where their wheat would flourish with the same promise of agricultural wealth. A marker from the National Register of Historic Places in the front yard of the Victorian house he built in Newton, Kansas, picks up the story:

"His letters home made him a leader of the Mennonite migration from Russia: about 5,000 Mennonites came to Kansas. . . . Warkentin operated several mills and promoted wheat growing, especially 'Turkey Red' hard winter wheat. His work helped make Kansas the 'Bread Basket of the World.'"

Indeed, the Great Plains would be infused with a new purpose: feeding the world. And so the Great Collision between nourishing and preserving on the American prairie intensified. Soon, the pendulum would swing heavily against preserving.

The newcomers rushing to stake a claim in the era of the Homestead Act couldn't foresee the consequences. In a matter of a few decades, intensive agriculture practices spread across the prairie land. Rising demand for wheat and corn domestically and in Europe during World War I pushed up prices for the commodities, prompting farmers to plow up millions of acres of native grassland. Even when the price of wheat and other crop commodities plunged after the Great War, farmers plowed up more land to compensate for the lower income so they could meet their loan payments. The planting of annual crops, often the same one in the same field year after year, replaced the diverse array of perennial grasses, flowers, and weeds that had provided constant ground cover. The annual plowing released carbon from the soil and began depleting the rich organic matter that had built up over millennia. With fewer deep-rooted grasses holding the soil in place, more and more land was left exposed to wind and rain between growing seasons, quickening the pace of erosion over time.

In the 1930s, the economic woe of the Great Depression spread across the country alongside an extreme climatic event: a prolonged drought that parched the soil of the southern Great Plains—from Texas to Nebraska, with the epicenter in the Oklahoma panhandle. Fierce, relentless winds lifted up huge roiling clouds, creating what were called "black blizzards." The vast dust storms buried crops and livestock and houses and barns, making farming nearly impossible for several years. Livelihoods, and lives, were wiped out. There was a gathering fear that the Plains were becoming a vast desert, like Africa's Sahara.

The emerging breadbasket became the Dust Bowl. The drought and the wind spread the precious topsoil of the Great Plains over the eastern half of the country; some of it even settled on the roof of the White House and the Capitol dome in Washington, D.C., and obscured the Statue of Liberty in New York City. Studies of the Dust Bowl's damage estimate that more than one billion tons of soil were lost across one hundred million acres of the Great Plains between 1934 and 1935, when the drought was at its worst. It was a harsh repudiation of the farmers' belief that their rich soil was a renewable resource.

The dust in D.C. and the devastation on the Plains alarmed politicians. In 1935, as part of President Franklin Roosevelt's New Deal, Congress established the Soil Erosion Service and the Prairie States Forestry Project. Under these programs, more than two hundred million trees and shrubs were planted on farms as windbreaks from North Dakota to Texas. It was the first widespread use of agroforestry, the practice now being deployed to regreen Africa. The feature of the Kansas Plains that Warkentin so admired during his search for future agriculture prospects—few trees!—turned out to be a big environmental drawback. A new federal agency, the Soil Conservation Service, encouraged farmers to use new methods like contour plowing and planting cover crops in between wheat seasons to hold back erosion. It also restored some land to prairie grass conditions and offered to pay farmers not to plant and let their soils rest.

Still, the expansion of agriculture on the Great Plains continued, especially as commodity prices rose again during World War II and the post-Depression economic recovery. Even though many farmers left the Plains after the Dust Bowl, and the number of family farms declined over the years, the number of acres under the plow continued to expand with the growth of large-scale corporate farming. The drive to push crop yields to ever-higher levels led to thicker application of chemical fertilizers and insecticides and pesticides. Rather than depend on rain from above, water was pulled up from below, from the Ogallala Aquifer that lies under much of the Great Plains. Through agriculture irrigation and household and commercial consumption, the aquifer is being drained faster than it is being naturally replenished by rain and melting snow. The U.S. Global Change Research Program, in its 2018 *National Climate Assessment* report, concluded that "major portions of the Ogallala Aquifer should now be considered a nonrenewable resource."

This is the wisdom of the prairie that Jason Schmidt understands: The soil and the water aren't limitless. They will vanish, if we let them.

But are we listening?

Between 2016 and 2020, the Great Plains lost more than seven million acres of previously intact grassland to crop production, according to the World Wildlife Fund's *Plowprint Report*; wheat, corn, and soybeans covered about 60 percent of the total planted acreage. The WWF estimated that just over half of the original native grasslands of the Great Plains were still intact. As the expansion of cropland continued, so did the erosion. Kansas, for instance, was losing an estimated 190 million tons of topsoil each year. The Union of Concerned Scientists, monitoring the impacts of environmental threats and climate change, warned in a 2020 study: "If soil continues to erode at current rates, U.S. farmers could lose a half-inch of topsoil by 2035—more than eight times the amount of topsoil lost during the Dust Bowl. They could lose nearly three inches by 2100. Given that it takes a century or more for an inch of soil to form naturally, the United States will lose the equivalent of at least 300 years' worth of soil by 2100 if today's trends prevail."

Back on his farm, Brandon Kaufman says, with a dose of gallows humor, "Our number one export in the U.S. per ton is topsoil."

◈

"We're in a time of huge transition," Brandon says, one that he hopes will focus on soil health and shift Great Plains farming away from increasing yields at all costs to preserving land to save costs. He drives out to another part of his farm, passing his neighbors' toasted wheat fields, and pulls up to a plot with a fresh green blanket where his transition began. It is a patch of Kernza, the ancient wild prairie grain that plant scientists and farmers like Brandon are working to develop into a commercial grain that can feed both people and livestock. It is a perennial that grows productively for several years after planting. Brandon sowed this field three years ago and hasn't touched the soil since, letting Kernza's remarkable root system expand. The bushy roots can burrow ten feet or more beneath the surface (three times deeper than the roots of annual wheat), delivering atmospheric carbon captured by the aboveground plant to the soil. They distribute nutrients to, and pull up water from, a wide underground area. His cattle grazed the Kernza for limited periods, fertilizing the soil with the nutrients in their manure. The only time Brandon disturbed the soil after planting was to dig out a small sample with the pliers hanging from his belt.

The results of that sample were astonishing. "I raised the organic matter 1.2 percent in three years, when people said that can't be done," he says. "But you get animal activity out there, and infiltration [of water and nutrients], and you have the soil covered all the time" to prevent erosion. "Look at the armor here," he notes with pride, indicating the tall perennial plants that have flourished during the drought, protecting and nourishing the soil.

He took another sample from a field on a hill that he planted with Kernza after years of diminishing yields from corn and soybean crops. "I remember my great-grandpa plowing the earth and it needed terraces so it wouldn't be eroding down the hill," Brandon says. To restore the deteriorating soil, he planted Kernza and grazed his cows in the field. At the time of the Kernza planting, he recalls, "there were no worms to be had there. You couldn't find an earthworm for nothing. Well, that first year I grazed it twice and I take my kids out there and with my pliers we dig up two inches deep and we find thirteen worms! Under the cowpies. So we created an environment for the worms, which is as good as having fertilizer out there. If you have a worm cycle, you get more soil, it's a worm slurry." In subsequent years he noticed bigger worms, and a greater variety of underground life. "I'm pretty sure that's a good thing. Huge worms. We've created a root system that is more beneficial."

Even more exciting to him than the earthworm proliferation was another insect discovery. "I've seen a couple of dung beetles the other day," Brandon says, a rare sighting on the central Kansas plains, due to heavy insecticide use. Those beetles roll dung into tiny balls (huge in comparison to the size of the beetles) and bury them in the soil to help with decomposition, making them very industrious fertilizer machines.

Brandon welcomed the beetles as his new allies in creating healthy soils on his insecticide-free fields. He shared the thrilling news with neighboring farmers, who were more alarmed than impressed. Beetles? In your fields?

Why do you want *them*? they asked.

"It's a sign I'm doing things right," Brandon told them. "It's a huge pat on the back!"

Patience is a virtue in agriculture, as crops take their time to mature. But there was a palpable sense of urgency in Brandon's fields. He was seeing more and more family farms, which were handed down through generations from the original homesteaders with their 160-acre parcels, now passing into the hands of larger agricultural operations. The costs of equipment, labor, and investments in the essential elements of farming were driving further consolidation that accelerated during the farm crisis of the 1980s, when mountains of debt, high interest rates, plunging land values, and escalating operating costs forced thousands of farmers into bankruptcy.

"My mom always told me you can do anything you want except for farming," he says, chuckling. "Well, here I am. But she went through the eighties." After a few years of playing in various football leagues after college, and then farming for others and picking up construction work, he came back to the family farm. When he did, he noticed that the farming landscape was radically altered.

"I go to western Kansas and it just baffles me out there. There's fifteen, twenty-thousand-acre farms," Brandon says. "As that land changes hands, it's going into hands that probably aren't going to allow it to change again."

He worries that the bigger the farm, the smaller the willingness to change. These farms, he fears, are so wed to annual crops like wheat, corn, and soybeans—and encouraged by government subsidies to stay in that marriage—that embracing perennials to improve soil health and regenerate the prairie lands will be a difficult switch. With the financial pressures bearing down on today's farmers, it will be tough to move away from a model of prioritizing ever-higher production and think about preserving as well as nourishing.

Instead, it has been innovative family farmers, along with Indigenous tribal communities, leading the way back to the balanced rhythms of the prairie. Like the Lakota nation in South Dakota, who are reclaiming their traditional farming methods and foods and reestablishing the once-mighty buffalo herd through its food sovereignty initiative. Like rancher Gabe Brown in North Dakota, who abandoned many of the accepted practices of conventional industrial agriculture and adopted regenerative cropping and grazing methods that focused first on rebuilding the soil to save his degraded and struggling farm, a journey he described in the book *Dirt to Soil*. Like farmer Jay Hill in Texas, who introduced water conservation methods such as drip irrigation to bring parched, unproductive scrubland to life with lettuce, onions, chili peppers, and alfalfa, and a regimen of forage cover crops like oats, sorghum, and rye grass to build up organic matter, capture carbon, and protect the soil from erosion. And like dairyman Jason Schmidt on his Grazing Plains Farm in central Kansas, who orchestrated the grazing patterns of his cows to frequently move from one patch of cover crops to the next, much like the buffalo once ranged on the prairies. He hopes that by maintaining healthy carbon-capturing grasslands with a polyculture of various perennial crops, he can somewhat offset the methane emissions of his cows.

"For me, I love livestock, I love grazing, so I look toward the prairies as to how to model how I graze my cows," he says. "I dream of how do we create agriculture ecosystems that are carbon sinks, that I'm guessing will have a lot more positive ramifications for soil health, for human health, for animal health, rather than a system based on monoculture crops and confined animals."

But getting there, moving from carbon emitter to carbon capturer, he acknowledges, isn't easy. "We're stuck in this paradigm of battling nature, which is only going to increase with climate change, when we really should shift 180 degrees. I'm thinking of this farm as a ship, when it's probably a little bitty boat in this ocean of agriculture. I want to turn this ship, I want to turn it away from the agriculture paradigm of farming against nature toward farming with nature. Man, I can't do it when I really, really want to."

He looks up from the picnic table, toward the barn, and notices his diesel gas tank, which keeps his equipment running with greenhouse-gas-emitting fossil fuel. He shakes his head and laughs wearily. "The irony is sitting right next to us," he says. "I'm right next to our fuel tank and we're talking about climate change."

Then he points to a grain silo and a long-ago inscription, still clearly legible in barely faded white paint. RFS 1946. His grandfather's initials, still standing the test of time after seventy-seven years. Jason himself left the farm to get his undergraduate degree in international development, with a focus on peace and justice, and then spent several years in volunteer service with the Mennonite Church. He came back to the family farm in 2011 and noticed a dramatic decline in the number of dairy farms in his county, from about 120 in his grandfather's day to just "two of us now." To survive, he knew he needed to turn the ship, and turn it fast.

As he and his dogs corral seventy-some cows from pasture to milking barn (cows that produce about three times more milk daily than Pasqualine's Kenyan breeds), Jason returns to his existential question: what is the wisdom of the prairies? "Well," he ventures, "it's perennial-based, it's polycultures, it's movement of herbivores. It's not disturbing the soil. What did that prairie provide? It was a protective blanket on the land. So I feel that's where agriculture on the Great Plains needs to look."

In its search for solutions, the U.S. Department of Agriculture (USDA), during the Biden administration, allocated about $3 billion to a program designed to reduce the agriculture sector's greenhouse gas emissions. The funding included projects that would encourage farmers to adopt practices such as no-till planting, sowing cover crops, and improving organic fertilizer and manure usage. As the program was announced, the Center on Global Food and Agriculture at the Chicago Council on Global Affairs released a report noting there was "substantial room to increase" such regenerative agriculture practices in the U.S. The USDA's 2017 Census of Agriculture found that less than 4 percent of all U.S. cropland had been planted in cover crops that year, and only 32 percent of U.S. cattle operations used intensive rotation grazing. The Council report noted that reduced or no-till farming was still not viewed by farmers as standard practice on U.S. fields.

Jason wasn't surprised that regenerative agriculture was having trouble gaining momentum in the Great Plains and across the country. Given the financial pressures on farmers, from the price of land to the cost of equipment, Jason says, "What's the first thing to *not* prioritize? It's environment and nature."

◈

Brandon Kaufman admires his fields blanketed with his polyculture of cover crops. In the past, he notes, farmers would brag about who had the highest yields, the straightest rows of crops, or the cleanest fields with the fewest weeds—all accomplishments requiring more technology and chemical enhancements. But rare were farmers like him, who would brag about things you couldn't see, about soil organic matter content or earthworm and dung beetle populations.

"I leave my weeds grow," he says defiantly. "Others might say it's a dirty field." But he would tell them the land is covered, the soil protected. It is recovering lost value, appreciating as an asset. He would ask them, "Does your banker ever ask, 'How clean is your field?'"

Brandon is betting that the defining question asked of farmers in the future will be: how deep is your soil organic matter? "I feel that in my lifetime, I'm going to see fields that are worth something and fields that are wasteland. If I can say I have four and a half percent organic matter, having started below 2 percent, now you've got acreage you can go back to. We've become incredibly tied to the monocrop system. But as people come to understand what's going on, as people start to see the disparity in land values . . . "

He pauses in thought. "If I sold some of the land that I own that has four to 6 percent organic matter," he continues, "people that get it are going to pay a huge premium for that. You can't build organic matter overnight. You can go buy phosphorus, potassium and dump all that stuff on it, but . . . "

He pauses again, and finally echoes Abebe's question: Why would you do that?

"I've got some fields that haven't been tilled in forever, like seventy years, and they're over 6 percent organic matter," he notes. "And in that percent of organic matter, there's a lot of phosphorus, there's a lot of water-holding capacity, tremendous amounts of nitrogen. So if we're taking fields and allowing them to appreciate by building soil health, maybe we're not getting cash payments today [from crop production] but your net worth, your asset, your portfolio, is growing by not depreciating, by not devaluating, with bio-management practices."

One of his greatest allies in his soil health movement is his herd of about seventy cattle. "They have four legs. I let them do the work," he insists. Like Gabe Brown and Jason Schmidt, he practices controlled rotational grazing, moving his cows around, letting them eat the forage cover crops growing on the fields. With this approach, the cattle are steered to another patch before they can graze down to the ground, leaving some green covering behind. All the while, they are also directly enriching the soil with their manure, full of the nutrients from the forage. Brandon calls it the "complete poop loop": forage grows in a field, cattle graze it, cattle poop on it, another crop grows.

Brandon first discovered that virtuous circle as a boy, working on his grandfather's farm. "He had a chicken and turkey shed," Brandon remembers. The poultry would feed on grains from the field, they would poop, "and we'd come through and shovel it and put it back in the fields. The fields are now very high in nutrients because of the chickens and turkeys."

Years later, he encountered the opposite: the incomplete poop loop, one of the follies of modern agriculture. He was working on a larger agriculture enterprise, where hundreds of cattle were being fattened in a confined feeding operation. They were being fed grain forage brought in from distant fields—forage that had been planted with expensive equipment, nurtured with water and chemical fertilizers, harvested with other expensive equipment, and then hauled off to feed the cattle. Brandon would collect the manure left behind in the pens and transport it back out to another field to fertilize another season of forage crops. "We were hauling manure and I hated hauling manure. It was just a load of turds after another load of turds," he recalls. "It's hotter than Hades, the flies are bad, and it just hits me, this is so stupid." It galvanized his determination to put his cattle to work directly fertilizing his fields, eliminating the middleman—him—hauling the forage and the manure from and to the same field, and instead replicating the prairie grazing of yore.

The main beneficiaries of this complete loop are Brandon's two true passions: polycultures and perennials. He abhors barren, brown fields; something is always growing on his land throughout the year. As he began shifting away from wheat and other annual crops in a monocrop routine, he planted a mixture of crops in his fields: sunflowers, milo, and rye to hold the soil in place; buckwheat, which helps other plants take up phosphorus in the soil; mung beans, which fix nitrogen; and Kernza, which delivers the nutrients deep into the ground with its epic root system. The thick green cover does a reliable job of suppressing weeds, depriving them of the sunlight they need to sprout. If some weeds do poke through, he welcomes them too.

"So we've got our ground cover, we've got our nitrogen, our phosphorus," he says. In a system like that, he figures, there is little need to buy fertilizer and herbicides. "I'm just buying seed and I'm paying myself to seed it. That can be pretty lucrative compared to what my neighbor is doing: $90 an acre in nitrogen, $60 an acre in herbicides, $25 an acre in insecticides. I did it all with $40 of seed. Now, the problem with that is you've got to get that established and get it going."

That's where the perennials come in. "With the rainfall patterns now, it's harder and harder every year to get something established just because we have windows that are very small, and when it does rain it's really intense, heavy rains. The reality of it is that the rainfall patterns are so inconsistent."

But he calculated that once the perennial crops were established, the revenue would grow as the plants did, season after season without interruption, and the expenses would shrink. "The great thing about a perennial is that once it's established, you've paid it forward for a number of years," Brandon says while wading into one of his Kernza fields. "And I'm not out here managing and seeding and buying seed every year, paying for equipment and burning fuel every year to try again. And with a perennial, you can grow it as a grain crop or you can use it as a forage crop, grazing the cattle, which to me is sort of a safety net. It's not either-or. If you graze in the winter or the fall you can also take a grain crop off in the spring. You're not just relying on grain. You're getting grazing income. You're increasing your soil health. How do you put a number to that?"

Here, in this field, was the Holy Grail of perennials: a crop that can feed both humans and livestock, a crop that can potentially be as prolific as wheat, nourishing people around the world, ubiquitous on tables everywhere, but only needs to be planted every three or four years. And it would keep growing season after season, and restoring the soil as it does. That's why Brandon had focused on Kernza. By the summer of 2023, he had cultivated about two hundred acres of it, making him one of the largest farmers of the ancient wheatgrass in the U.S.

"Kernza will stay in the field until I terminate it. For grain production, three years is about your window. But I've paid it forward. I'm going to graze it hard with cattle. If I can get four high-quality grazing periods out of it, and take a grain crop from it . . . "

Once again, he pauses in thought, this time pondering a future of perennial food crops flourishing across the prairie. "I don't know if Kernza is the perennial grain that takes off," he says. "But there will be one that does."

◈

Lee DeHaan emerged from another Kernza field less than an hour's drive from Brandon's patch, straight up highway 135 in central Kansas. Lee is the director of Crop Improvement and the lead scientist of the Kernza Domestication Program at the Land Institute just outside the city of Salina. Finding that Holy Grail of a perennial grain that can both nourish and preserve the planet has been his lifetime quest.

As a boy growing up on a family farm in southern Minnesota, Lee delighted in discussions with his father about the challenges of farming. "My father was a forward-thinking farmer who saw the issues with agriculture and tried to do everything with that system to build healthier soil and protect the land," Lee recalls. One day in the early 1980s, his father heard a presentation by Wes Jackson, a plant geneticist who had founded the Land Institute with his wife, Dana, several years earlier. The farm crisis was ravaging the American Midwest. As commodity prices dropped and profits shriveled and debts piled up, farmers struggled to hold on to their land. At the same time, the health of the very soil they depended on was declining through erosion and depletion of organic matter. The government devised a conservation program that would pay farmers to stop planting corn and soybeans and instead plant grasses to rebuild the soil and heal the land. For Lee, just beginning to ponder the economics of farming, it was a head-scratching dilemma. How do you do three things at once: produce food, conserve the environment, and preserve farmer livelihoods?

"We saw this tradeoff between the ability to grow lots of food in an intensive corn-soybean system with the downsides of environmental consequences and even a system that wasn't allowing farmers to be profitable. We had great production, then we're pulling these lands out of production and really conserving the soil, but now we don't have food production, we don't have farmer livelihoods." Lee could see how these tradeoffs were vexing his father. "My father's great joy in his work was always this idea that 'I'm growing food for people.' That's a very rewarding calling to be feeding people and be the person responsible for growing food." But how do you reconcile that with also being responsible for damaging the environment?

"Then," Lee says, picking up the story of their eureka moment, "my father heard Wes Jackson talk about the idea of perennial grains. What we soon realized was that this was a way to meet all three of those things at the same time with one solution." This potential stirred the budding farmer's imagination. "If we could instead plant a perennial grain that grows like the prairie grass, builds the soil, protects the soil, but also yields food we eat. If we could use less pesticides, not till the soil all the time, and reduce costs to the farmer. That solves all three things at once. It seems like a brilliant solution."

But all the major grains—wheat, corn, rice—were annuals, needing to be replanted every year after just one harvest. Where was this perennial grain seed? Lee set off to study the possibilities. At the University of Minnesota, he earned graduate degrees in agronomy and crop science, and he became an early student of what would become the Forever Green Initiative led by Don Wyse. Research on perennials and other crops that would both nourish and preserve was sprouting on the same campus trial plots where Norman Borlaug did early work that led to the Green Revolution. The new challenge was more complex: still increasing yields, yes, but also now limiting or eliminating the environmental consequences as well.

All the while, Wes Jackson's vision to develop perennial grain crops was attracting new disciples. Plant breeders at the Rodale Institute in New York had selected a grass that had long grown on the steppe of Asia and Europe called intermediate wheatgrass, a species related to wheat, as a promising perennial candidate. For several thousand years, the grass had been used as forage for livestock, particularly in its native region of western Asia; in the 1930s, the USDA introduced it to the U.S. as a forage crop. A decade earlier, Russian scientists had begun researching the perennial's potential, but that work was ended with revolution and the rise of the Soviet Union. Further research was largely dormant until the 1980s, when the Rodale breeders began working with the wheatgrass on various traits including nutrition, harvestability, and food-use potential. With seeds from that program, the Land Institute began its intermediate wheatgrass breeding trials in 2003, and in 2009 it filed for the trademark name "Kernza," a combination of the words "kernel" and "Konza" from the Konza Prairie that was the tribal homeland of the Kaw Nation. The next year, a farm in Kansas was the scene of the first large-scale harvest of Kernza, producing a semi-truckload of grain.

Over the following decade, Kernza research and production expanded to multiple countries, including Australia, Canada, France, Sweden, Ukraine, Uruguay, and the United Kingdom. In the U.S., Kernza was growing on about four thousand acres, primarily in Kansas, Montana, and Minnesota. In the labs at the Land Institute's busy campus in Salina, on the banks of the Smoky Hill River, scientists worked to increase the seed and grain size of Kernza to approach the scale of annual wheat, which would improve the bread-baking quality. Their ultimate ambition was to develop a Kernza strain that would have yields similar to annual wheat and could flourish in fields around the world.

For that to happen, the grain's commercial appeal would need to match the agricultural progress, creating a demand from consumers as it moved from farm to market to plates. "The issue isn't finding farmers to grow it, but finding consistent markets," notes Peter Miller, joining Brandon Kaufman in his Kernza field. The two had teamed up along with Land Institute perennial program scientist Brandon Schlautman to form Sustain-A-Grain, an enterprise that would introduce consumers to Kernza and support family farmers growing it. Peter had grown up on the Kansas plains and was working in food and agriculture startups when he met Brandon at a college class reunion and heard all about his ambition for Kernza. For Peter, it brought back memories of a long-ago muffin he had sampled at one of the Land Institute's Prairie Festivals, with a sweet, nutty flavor made from prairie grass seeds. It was a bite that introduced the notion that there were edible grasses and grains out there other than the crops grown on his family farm. Hearing years later that Brandon was now growing such a tantalizing prairie grain inspired Peter to spread the word about its potential. If only everyone could taste it.

Kernza, which behaves similarly to wheat during baking and cooking, with a higher protein content, had first appeared on a restaurant menu at the Birchwood Café in Minneapolis in 2013: Kernza waffles topped with apple shallot compote, cinnamon honey butter, pepitas, bacon lardons, a sunny-side-up egg, and maple syrup. According to a Land Institute Kernza timeline, new recipes appeared in short order—Kernza pancakes, cookies, breads, noodles. In 2016, Patagonia Provisions, the food division of parent company Patagonia, launched the appropriately named Long Root Ale, the first packaged Kernza product on the market. Other brewers concocted additional Kernza beers. General Mills, which gave us Cheerios and Wheaties, committed to a trial development of Kernza cereal with Cascadian Farm in Washington's Skagit Valley, a pioneer in the movement to nourish and preserve. In 2020, Perennial Pantry debuted as the first dedicated Kernza processor and food brand, launching online Kernza flour sales.

As Peter and others worked to build traction for Kernza products in a price-sensitive market skittish of new ingredients, the perennial grain was gaining momentum on the environmental front. In 2016, the Minnesota Clean Water Council included Kernza in their annual policy recommendation as a cover crop that could help with drinking water protection. A number of Minnesota counties, particularly those with shallower water table wells, had been reporting rising levels of nitrates in their drinking water, resulting mainly from runoff of nitrogen fertilizer and other agrochemicals used to grow annual crops like corn, soybeans, and wheat. These annuals absorb only about 50 percent of the nitrogen that is applied as fertilizer, with the rest leaching into the environment. The Minnesota Pollution Control Agency reported that nearly half of wells in agricultural parts of the state had nitrate concentrations higher than the federal Environmental Protection Agency standards for drinking water. Consuming too much nitrate can affect how blood carries oxygen and can lead to blue baby syndrome, especially in bottle-fed babies younger than six months old. It can also harm young livestock. State and local officials embraced the Kernza-pioneering farmers as allies in protecting wellheads, as the long, bushy, beard-like roots were proving effective in intercepting chemicals and absorbing nitrates making their way down through the soil before they could enter the water system. Kernza became particularly popular in southwestern Minnesota counties bordering South Dakota and Iowa as a tool to reduce pressure on water treatment facilities; growing Kernza was far cheaper than spending millions to upgrade mechanical filters.

These environmental successes have given Peter and the purveyors of Kernza products a valuable marketing hook. Patagonia's fusilli boxtop trumpeted that Kernza "regenerates topsoil, draws down carbon," and encouraged consumers to "eat up to draw down." Perennial Pantry, on its Kernza flour packaging, noted, "Kernza isn't a 'less-bad' climate solution—it's actually good for the environment *and* farmers (and you!). Amazing, right? Let's get cooking and help save the planet."

Right there, on that label, were the three things that Wes Jackson envisioned and that had brought Lee to the Land Institute more than two decades earlier: a perennial crop that nourished people, preserved the environment, and enhanced farmer livelihoods. As Lee says, standing in his Kernza field, it seems like the long-sought solution. But he also acknowledges that "It's not an easy solution. It's a long-term solution. You have to have your sights on the end goal and have the funders, supporters who are willing to work toward developing these perennial grains over decades, and all the things involved in creating new crops."

But that support has been slower to develop than the painstaking efforts of breeding the crops. Borlaug's work to increase yields of the staple annual grains during the Green Revolution became a global priority, attracting billions of dollars in support from the largest philanthropic institutes at the time and from governments and heads of state all over the world, including the U.S. In contrast, the efforts to develop new perennial crops that can both nourish people and preserve the land has often felt as isolating as rural life on the Kansas plains.

The Land Institute, with a budget of about $11 million, received funding mostly from private individuals and philanthropies. In 2022, the Minnesota legislature approved a Continuous Living Cover Value Chain Development Fund to provide support to Kernza supply chain partners. It also provided long-term funding for the Forever Green Initiative at the University of Minnesota and its research on Kernza and other perennial crops. But funding from the federal government for research on perennials had been miniscule. The Land Institute calculated that in the decade from 2011 to 2021, the USDA invested more than $14 billion in agriculture research grants, with the vast majority of that going to conventional agriculture and annual crops. Of the total grant money, the portion that included the word "perennial" was just one-quarter of one percent.

This has been mystifying and frustrating to the scientists and farmers itching to do something different, particularly as the pace of extreme weather events and environmental disasters has quickened. "All of society has to come together around developing these new food systems that provide not just food but also all these environmental services that we all expect and hope to have from agriculture. Like clean water, soil conservation, soil sequestration of carbon, which will help address climate change while building soil quality at the same time," Lee DeHaan insists. "We see great potential to find a grain that can do that, if only we can develop it."

If only. "We haven't tried seriously yet as a society, to say, 'Yes, we're doing this,'" Lee adds. "We're not yet like the Green Revolution where people said we'll have starving en masse if we don't do this."

The urgency of the task is palpable at the Land Institute campus, where an array of potential perennial crops command the scientists' attention in a number of indoor labs and outdoor test plots. Lee estimates he has collected research from well over one hundred thousand plants, exploring the potential for breakthroughs. While the buzz around Kernza had been the most prominent, researchers also worked on the perennial possibilities for a range of plants and crops, like sorghum, wheat, rice, alfalfa, soybeans and other legumes, and oilseed crops like sunflowers and silphium. Much of this research involved collaborations with labs and institutes at universities throughout the U.S. and around the world. Meticulously, patiently, from test harvest to test harvest, the scientists narrow down traits that might be suitable for farmers to eventually cultivate and harvest on a large scale, for processors to be able turn into food products, for customers to be able to safely eat and enjoy.

The first question preoccupying the scientists is, Will it work in the field? The second question is, How does it taste? The Institute's staff double as the initial taste testers. "Those of us here who have eaten them are still alive," Lee jokes. "So that's a good data point."

In one of the labs, a big white bag of Kernza seeds freshly delivered by Brandon Kaufman awaited the ultimate test. They would soon be ground into flour and transformed into muffins, cookies, pancakes, breads, and other baked goods for education programs of the Land Institute to whet the public appetite for this new crop. While the 2023 wheat crop wilted in the drought, Brandon's Kernza flourished; in one field, he harvested four hundred pounds an acre, double the usual yield. It was an encouraging sign; if his Kernza could supplant his ancestors' hard red winter wheat in household cupboards, it would boost the possibility of his children continuing the stewardship of the family farm. "That's my legacy," he says. "Otherwise, what's the point?"

The thought of Kernza catching on in kitchens brings Brandon back to the soil, and the need to restore the rich organic matter that makes farming and nourishing possible. "It's like cooking," he notes. "If you run out of sugar, you're not making cookies."

That is also the unmistakable message at the Land Institute: it all starts with the soil. Illustrations and exhibits reveal the depletion of the Great Plains soil over time, from the onset of the "catastrophe of tillage" in the late nineteenth century. Within several decades after that, the soil lost 50 percent of its carbon stock. During that time, farmers grew wheat without fertilizer, relying on the carbon and nitrogen stored in the soil. As that depleted, ever-greater levels of chemical fertilizer were introduced, creating "the illusion of success" despite the soil degradation.

"It's a short blip of time that so much carbon has been lost. It was built over centuries," explains Aubrey Streit Krug, the director of the Institute's perennial cultures lab. "This is a crucial decade to make mitigation changes." Lee adds, "Good soil helps to mitigate all this climate volatility."

Across the road from the labs is a stretch of remnant

native prairie that still exists in its ancient state, never tilled or cultivated or disturbed by anything other than natural actions. Lee's nearby Kernza field is the showpiece of this diverse perennial ecosystem where many of the plants and crops in the labs grow under real prairie conditions. "We're seeing diversity of plants in this ecosystem that are competing and cooperating with wildlife," Aubrey says. It is the Kansas version of Kenya's Kapiti Ranch, where domesticated livestock roam beside Africa's wildlife. Here, Aubrey notes, "We can see how this has existed all this time. These grassland prairies persisted for hundreds of thousands of years. We are using that as a model and measurement."

She closes her eyes and inhales deeply. "Here," she says, "we can imagine what the Great Plains can look like again. It is a vision for people and places together, moving from being more exploitative to more responsible. As opposed to seeing these systems as needing to be replaced, as we have before, we're learning *with* natural systems as the Indigenous people did."

◈

A bit more than four hundred miles further north, a flat drive through the heart of the Great Plains to just across the Nebraska border into South Dakota, another patch of prairie land yearns to return to the way things were, before the wheat and plows arrived and the buffalo disappeared. Here, at the home of the Sičaŋǧu Lakota nation, a food sovereignty initiative is reviving the Indigenous principles of regenerative agriculture, passed down through generations and compiled in poetic form by today's storytellers:

"We are Sičaŋǧu
We see a world
Where we are healthy and whole

In body, mind, and spirit
Connected to the earth and stars
And all our relations
We are living that world into being

. . .

We were here before
We know the way
We persevere
We are Lakota
We are Siċaŋġu

Whenever anyone in North America uses the phrase "regenerative agriculture" or talks about working with nature to nourish and preserve the planet, an acknowledgment must be made that the inspiration begins here, and on all the other Indigenous lands of the Great Plains and beyond.

We were here before
We know the way

So we end our journey through ancient lands in this place of beginnings. Where the buffalo, and genuine food sovereignty, are returning.

"Aren't they magnificent?" TJ Heinert asks as he admires a small group of the new arrivals grazing on prairie grass. He is one of the managers of the Wolakota Buffalo Range, where more than one thousand buffalo roam on Lakota land. "What we do as a nation we learned from the buffalo. We are the buffalo people, we are one. We say everything is related. For the Lakota people, everything is a relative. And it is so great to have these relatives back."

Here on this patch of the Plains, regenerative agriculture

goes beyond improving soil health, storing carbon, healing degraded lands, and nurturing "climate smart" crops. Here it is also about regenerating relationships, so that human, animal, and plant spirits can be restored as well.

Those natural relationships of the Great Plains were brutally severed during the nineteenth century military campaign against the Native American communities and the food sources that gave them life, health, and spiritual connection. Their land was stolen and restricted, their food sovereignty taken away, their few remaining buffalo fenced into new national parks. In the many decades since, these Indigenous communities of the Great Plains and throughout the U.S. have become some of the poorest, most food-insecure, least healthy places in America. Their land, once abundant, devolved into food deserts with a dearth of access to fresh foods; the residents, once robust with diverse, nutritious diets, were forced to rely on poor-quality treaty rations provided by the army that had driven them off their land and then on the cheaper high-calorie, low-nutrient, microwave-ready processed foods pushed at them by the modern agriculture system. The Native American poverty rate here on the Rosebud reservation, and on most reservations, exceeded 60 percent, more than four times greater than the national average. Diabetes and obesity rates were among the highest levels in the country. According to the Sičaŋǧu Health Initiative, the life expectancy for men was forty-eight years, and for women fifty-eight years—appalling numbers in a country with an average life expectancy of about seventy-seven years.

"All of this traces back to food," says Jillian Waln, marketing and storytelling director for Sičaŋǧu Co and its 7Gen Vision, which imagines the world they would like to see their descendants living in 175 years from now. Like Cool Waters Nairobi and Delhi Urban Farms, 7Gen is one of the Rockefeller Foundation's 2050 Vision finalists—though the Sicaŋġu vision reaches much further into the future.

At the core of the 7Gen approach is a food sovereignty initiative to reestablish lost practices, plant new traditions, and reclaim kinship relationships with the land and environment. "If you want to completely subjugate and oppress another society, one of the first things you do is you control their food," community leader Wizipan Little Elk explained. "Beginning with the mass slaughter of the buffalo, about $2 trillion of wealth has been extracted from our people. . . . In order for us to regain our power, we have to regain our food."

Matte Wilson, the director of the food sovereignty initiative, framed the vision in a poignant essay published by the Thomson Reuters Foundation in early 2023:

"Before the military campaigns that sought to eliminate us, Indigenous Peoples thrived because our social and economic systems were based on relationships with nature and the environment. We prioritized these relationships, a concept we call *Wotakuye*: all life is treated as relatives. *Wotakuye* still informs Lakota food ways and agriculture, providing not only for nutrition and health, but also for education opportunities, employment, and strong communities. . . . This is how we must approach regenerative agriculture—not in silos, but as part of an integrated system that holds enormous promise as a solution to climate change, food insecurity, poor health, and the inequities of the industrialized food system."

His words are an eloquent manifesto for Indigenous leadership in the global agriculture regenerative movement. He welcomes the efforts of multinational food and agribusiness corporations to respond to climate change disruptions to their supply chains and shifting consumer demands by lessening the environmental impact of their business operations. But, Matte cautions, "If we want to unlock the power of regenerative agriculture, we need to embrace a deeper meaning and look for leadership elsewhere: starting with Indigenous Peoples. After all, Indigenous Peoples occupy, manage, and survive on around one-quarter of the Earth's land, which contains 80 percent of the world's remaining biodiversity—a sign that we are among the most effective stewards of the environment."

He continues:

> The current global food system produces the diets that account for one in five deaths worldwide. It generates a third of all greenhouse gas emissions, destroys biodiversity, and leaves too many agricultural workers and their families in poverty. It would be a grave mistake to trust the same corporations and corporate values that got us here to lead us out.
>
> Regenerative agriculture holds great promise—which is why Indigenous Peoples and farmers practicing it should be at the center of any serious agenda to better understand and scale it. Many of the biggest corporate programs that brand themselves as "regenerative" focus solely on improving soil health and storing carbon. But this is only one small part of what is needed to make agriculture regenerative of the earth, and of human and animal health. . . .
>
> Three things are needed to realize this promise. First is to amplify Indigenous voices and the work being done in our communities. Second is to invest in that work. Organizations and societies that have benefited from the wealth extracted from Indigenous lands and cultures have a duty to make amends. Third is to recon-

figure the global, national and local advisory and planning bodies that are hammering out a course of action on regenerative agriculture.

Indigenous Peoples must be at the table not only as equals, but as leaders. The views and interests of a broad spectrum of farmers, land stewards, non-profits, communities, businesses and agencies must also be represented. Corporations will also be there, but they cannot dominate. Otherwise, we risk reducing regenerative agriculture to a marketing label, rather than a unified effort to transform our food systems to be healthy, resilient, sustainable, and prolific.

Deeply regenerative agriculture can only be achieved with Indigenous Peoples as equal partners in its definition and implementation. And only change of that depth can restore balance to the climate and health to the planet.

The Siċaŋġu Food Sovereignty Initiative began with a community farm and two greenhouses. One is shaped like a geodesic dome, where a range of foods scarce on the reservation grow, shielded from the harsh conditions of the windswept northern Plains: bananas, broccoli, squash, tomatoes, lettuce, cabbage, figs, lemons, grapes, cucumbers, kale, kohlrabi, onions, and herbs.

Bananas, lemons, and figs in South Dakota? "It is to show what is possible," Matte says with a smile.

The initiative designed a mobile market to bring this produce to communities far from any grocery store. The greenhouse doubles as a gathering place for young and old alike to learn about nutrition and the value of food, and the ways to harvest wild foods. An internship program sprouted to encourage youth to become food entrepreneurs, called "We will grow into producers."

"We want to get people excited about this, to educate our younger kids on what's out there, what can you do to help your tribe find their way back to when we were prosperous, when we were at peace," says garden manager Michelle Haukaas. What began as a one-acre trial plot soon expanded to six acres. "We want to get them onto the land, get as many tribal producers as possible. Here, we're so limited in our food choices. When you have to hustle for food, you can't really be at peace and try to find your purpose. It creates a lot of anxiety. 'What am I going to eat today, how am I going to find food today?'"

The central garden has inspired dozens of growing projects throughout the community and has linked up with regenerative agriculture efforts on other reservations. "I feel there's a lot of new, innovative, creative Natives out there who are doing it, and they're really inspirational," Michelle says. "We're just trying to find our way back to our old thought processes, bring back our kinship and values."

Restoring the land, she notes, has certainly restored her. "This garden really changed my life. It took my hopelessness away from me," she says. "It is our first step, to open people's eyes, reconnect with the land."

Regenerating connections is also a focus of the Lakota Immersion School, which began with kindergarten to second-grade children and plans for expansion to fifth grade. "We'll keep building, based off what the community wants," says Sage Fast Dog, the school director, who emphasized reestablishing the Lakota language and reasserting Indigenous identities. "The purpose of founding the school was to inspire and show others that we don't have to accept things the way they are."

He points to a mural painted on an inside wall depicting

the impact of the boarding school system imposed after the military campaign, which, he says, erased the children's tribal identity, shamed them for their culture, and forced them to become "uniform people. They came into school as tribal kids, they knew how to hunt, they knew the land, how to sustain life in the environment."

A central part of the curriculum highlights the community's annual buffalo harvest. The children learn about the history and vital importance of the buffalo. At the harvest, where an older male buffalo is culled from the herd, the community gathers for the sharing of every bit of the animal. They learn what are the edible pieces, what is used for tools and clothing, and what are the spiritual elements, like the skull. "They see that nothing is wasted," Sage Fast Dog says. "They learn, only take what you need."

It is a fundamental lesson of regenerative agriculture: the land, if protected and preserved, will provide all you need for an abundant life. "That's what we want the students to learn," he adds. "The importance of land, the stability of land."

At the center of the Lakota land is perhaps the most ambitious project of the food sovereignty initiative. In two years, the Wolakota regenerative buffalo range brought back more than one thousand buffalo to roam on twenty-eight thousand acres, making it the largest Native-owned herd. The goal, says TJ Heinert, is to revitalize not only the prairie but also the people who call it home.

"Having the buffalo back here on Native lands, to bring back these relatives, is huge for us. To be able to see them thrive on their homeland, where they belong, is really tremendous to see. They're healthy, happy, and they're home. We have so much to learn from them. In just two years, you can see the impact on the land, you can see the pastures where the buffalo grazed last year, and you see that regrowth is happening. To bring them back and recreate the ecosystem is an amazing thing."

He continues: "It's not just for us now. We're creating that ecosystem and bringing these relatives back for a greater purpose. To be a stepping stone for other tribes and other people all around the world to be able to see what we're doing in a small reservation here in South Dakota. . . and say, 'Hey, if they're doing it, so can we.'"

The buffalo arrived at Wolakota from state and national parks and other reserves across the Great Plains and the West. The births of new calves, TJ hopes, will steadily build the numbers to about 1,500, a herd that could roam, and teach, far into the future. "We talk about the 7Gen approach," he says, "so bringing these animals back is recreating our ecosystem to have something better for the next generations."

Every day, TJ patrols the range, gathering an intimate knowledge of the contours and residents of the land. "The buffalo were just here, grazing for about one and a half to two months," he says, his all-terrain vehicle rolling to a stop on a hilly patch blanketed with short prairie grass. "Now they have moved on and this area will get a year's worth of rest." By the time they return, he says, the grass will be waist-high again.

TJ marvels at the buffaloes' instinct for managing their environment. "The beauty of these animals is they don't graze every living thing. They leave plants for other animals." He laughs, thinking about an old tune. "It's why the song says, 'Where the buffalo roam!' Not where the buffalo stand and eat."

He continues the drive, pulling up beside a group of domestic cattle grazing on the other side of the fence, on private ranchland. "The cows, they just stand and eat, mowing the grass down to the soil and the sand," he says. "There will be nothing left when they move on. When the cattle over-graze, an entire pasture will be reduced to the sandy soil."

He often spots buffalo and cattle grazing across from each other, with a fence in between. TJ, who is also a rodeo cowboy, imagines the buffalo watching the cattle eat, munching the grass down to the nub, and asking each other: Now why would they do that?

It was the same fundamental question of regenerative agriculture asked by the farmers of the Great Rift Valley, and the Indo-Gangetic Plain, and the Pan-American highlands, and the Kansas Plains, about some of their conventional agriculture practices that harmed the environment. Why do we do that?

TJ considers buffalo to be the perfect agent of regenerative agriculture, easing the Great Collision between nourishing and preserving. Their grazing method inspired the new practices pioneered by the regenerative ranchers and dairy farmers who move their animals from pasture to pasture while the grasses and other forage plants still have several inches of growth remaining. TJ had heard, in stories passed down by Lakota ancestors, that wherever the buffalo went, other animals followed. Now he witnessed that himself at Wolakota. As the buffalo herd increased, so too, did the deer, elk, and antelope populations.

"The wildlife *do* follow the buffalo, they clean up what's left behind," he says. "They are bringing back different plant life, too." As they move and graze and poop, the animals spread seeds for new grasses and forage plants. TJ notes that the shape of buffalo hooves allow them to knead the soil differently than cattle, aerating rather than trampling.

"We have seen sage come back. Plants that are our cultural medicines are coming back. There's new grasses and different insects. All of these benefits, because we're bringing our relatives home. Having these relatives back on the land is so good for the ecosystem."

The buffalo, he says, have also "brought back our diet. They have brought back the proteins taken from us. The meat is leaner, less fat. We utilize all the organs in our cultural cuisine. We make broth from the bones. Every inch of the animal is used."

As an autumn sun sets on the Wolakota range, TJ shuts down the ATV and walks through two-foot-high prairie grass, toward a gathering of about a dozen buffalo. They continue grazing, undisturbed. They know him, and he knows them. "To witness their greatness is such a privilege," he says. "It is amazing to bring these iconic animals back on Native land."

The air is still and silent, the only sound coming from the chewing of grass and the pawing of ground. An October chill hints at the winter to come. Here, now, after so many decades of upheaval, the prairie of the American Great Plains is at rest—healing, regenerating.

"You can tell," TJ says, "the land itself is happy."

NOTES ON SOURCES

THE NARRATIVES OF THE farmers and their efforts to nourish their families and communities while also preserving their land and environment are from my own on-the-ground reporting. Passages on the wider context of their efforts, as well as agriculture's contribution to food security and its impact on the planet, are based on my interviews with sources named in the text and from research resources, including the following:

PROLOGUE

"African Forest Landscape Restoration Initiative (AFR100)." World Resources Institute.
wri.org/initiatives/african-forest-landscape-restoration-initiative-afr100

Almond, Rosamunde, Monique Grooten, Diego Juffe Bignoli, and Tanya Petersen, eds. *Living Planet Report 2022: Building a Nature-Positive Society*. Gland, Switzerland: WWF, 2022.
wwfint.awsassets.panda.org/downloads/embargo_13_10_2022_lpr_2022_full_report_single_page_1.pdf

Daskalova, Gergana N., and Johannes Kamp. "Abandoning Land Transforms Biodiversity: Land Abandonment Is Critical When Assessing Global Biodiversity and Conservation." *Science* 380, no. 6645 (May 2023): 581–3.
science.org/doi/10.1126/science.adf1099

FAO, IFAD, UNICEF, WFP, and WHO. *The State of Food Security and Nutrition in the World 2022: Repurposing Food and Agricultural Policies to Make Healthy Diets More Affordable*. Rome: FAO, 2022.
fao.org/3/cc0639en/cc0639en.pdf

Foley, Jonathan A., Navin Ramankutty, Kate A. Brauman, et al. "Solutions for a Cultivated Planet." *Nature* 478 (2011): 337–342.
nature.com/articles/nature10452

GAP Initiative. *Global Agricultural Productivity Report 2021: Strengthening the Climate for Sustainable Agricultural Growth*. Virginia Tech College of Agriculture and Life Sciences. November 17, 2021.

Left: *Spreading the Word*
New ideas carried on old conveyances

cals.vt.edu/global/cals-global-news/202111118-newsletter/2021-GAP-Launch.html

"Global Climate Highlights 2023." Copernicus Climate Change Service. climate.copernicus.eu/global-climate-highlights-2023

"Goal 2: Zero Hunger." United Nations. un.org/sustainabledevelopment/hunger/

Haberl, Helmut, Cheikh Mbow, Xiangzheng Deng, Elena G. Irwin, Suzi Kerr, Tobias Kuemmerle, Ole Mertz, Patrick Meyfroidt, and B. L. Turner II. "Finite Land Resources and Competition." In *Rethinking Global Land Use in an Urban Era*, edited by Karen C. Seto and Anette Reenberg, 35–69. Cambridge, MA: MIT Press, 2014. curis.ku.dk/portal/files/108467792/2014_ESF_land_use_competition_Haberl_et_al.pdf

Hanson, Craig, and Janet Ranganathan. "How to Manage the Global Land Squeeze? Produce, Protect, Reduce, Restore." World Resources Institute, 2023. wri.org/insights/manage-global-land-squeeze-produce-protect-reduce-restore

Ivanovich, Catherine C., Tianyi Sun, Doria R. Gordon, and Ilissa B. Ocko. "Future Warming from Global Food Consumption." *Nature Climate Change* 13 (2023): 297–302. nature.com/articles/s41558-023-01605-8

Millennium Ecosystem Assessment. *Ecosystems and Human Well-Being: Synthesis*. Washington, DC: Island Press, 2005. millenniumassessment.org/documents/document.356.aspx.pdf

Pörtner, Hans-Otto, Debra C. Roberts, Melinda M. B. Tignor, Elvira Poloczanska, Katja Mintenbeck, Andrés Alegría, Marlies Craig, Stephanie Langsdorf, Sina Löschke, Vincent Möller, Andrew Okem, and Bardhyl Rama, eds. *Climate Change 2022: Impacts, Adaptation and Vulnerability; Contribution of Working Group II to the Sixth Assessment Report of the Intergovernmental Panel on Climate Change.* Cambridge University Press, 2022. doi.org/10.1017/9781009325844. report.ipcc.ch/ar6/wg2/IPCC_AR6_WGII_FullReport.pdf

Project Drawdown. drawdown.org

Searchinger, Tim. *World Resources Report: Creating a Sustainable Food Future.* World Resources Institute, 2018.

research.wri.org/sites/default/files/2019-07/creating-sustainable-food-future_2_5.pdf

Searchinger, Timothy, Liqing Peng, Jessica Zionts, and Richard Waite. *The Global Land Squeeze: Managing the Growing Competition for Land.* World Resources Institute, 2023.
files.wri.org/d8/s3fs-public/2023-07/the-global-land-squeeze-report.pdf?VersionId=edANDGIvq_NhCGbDVfte6diBdJswo7e9

Stanford University and Carnegie Institution. "Abandoned Farmlands Are Key to Sustainable Bioenergy." ScienceDaily. June 24, 2008.
sciencedaily.com/releases/2008/06/080623113722.htm

United Nations Department of Economic and Social Affairs, Population Division. *World Population Prospects 2022: Summary of Results.* UN, 2022.
un.org/development/desa/pd/sites/www.un.org.development.desa.pd/files/wpp2022_summary_of_results.pdf

CHAPTER 1

"24 billion tons of fertile land lost every year, warns UN chief on World Day to Combat Desertification." UN News. June 16, 2019.
news.un.org/en/story/2019/06/1040561

"Climate." CIFOR-ICRAF.
cifor-icraf.org/our-work/areas/climate

"Countries in the World by Population (2023)." Worldometer.
worldometers.info/world-population/population-by-country

Dhaliwal, Bani. "How Deforestation Affects the Water Cycle." EarthDay.org. February 8, 2023.
earthday.org/how-deforestation-affects-the-water-cycle

Government of Ethiopia. "Managing Environmental Resources to Enable Transitions to More Sustainable Livelihoods." August 21, 2011.
reliefweb.int/report/ethiopia/meret-project-sustain-over 15-mln-people-livelihoods

"Great Green Wall Initiative." UN Convention to Combat Desertification.
unccd.int/our-work/ggwi

Hoije, Katarina. "Africa Needs Up to $65 Billion Loans Yearly to Curb Food Imports." Bloomberg. January 24, 2023.
bloomberg.com/news/articles/2023-01-24/africa-needs-up-to-65-billion-loans-yearly-to-curb-food-imports#xj4y7vzkg

Hoornweg, Daniel, and Kevin Pope. "Population Predictions for the

World's Largest Cities in the 21st Century." *Environment and Urbanization* 29, 1 (2017): 195–216.
journals.sagepub.com/doi/10.1177/0956247816663557

Institute of Human Origins. "Lucy's Story." Arizona State University.
iho.asu.edu/about/lucys-story

Stallwood, Paige. "Desertification in Africa: Causes, Effects and Solutions." Earth.Org. December 15, 2022.
earth.org/desertification-in-africa

"The Field Report." International Fund for Agricultural Development.
ifad.org/thefieldreport

United Nations, Department of Economic and Social Affairs, Population Division. *World Population Prospects 2022: Data Sources.* UN, September 2022.
population.un.org/wpp/Publications/Files/WPP2022_Data_Sources.pdf

Wikipedia. "List of Countries by Population in 2000." Last modified July 26, 2023.
en.wikipedia.org/wiki/List_of_countries_by_population_in_2000

World Agroforestry.
worldagroforestry.org

World Vision. *Farmer Managed Natural Regeneration: A Holistic Approach to Sustainable Development.* World Vision, 2019.
wvi.org/sites/default/files/2019-12/FMNR%20Publication%203Dec_Online_0.pdf

CHAPTER 2

Iowa State University Center for Sustainable Rural Livelihoods.
csrl.cals.iastate.edu

Mutesi, Fatuma, John Robert Stephen Tabuti, and David Mfitumukiza. "Extent and Rate of Deforestation and Forest Degradation (1986–2016) in West Bugwe Central Forest Reserve, Uganda." *International Journal of Forestry Research* (2021).
downloads.hindawi.com/journals/ijfr/2021/8860643.pdf

Reynolds, Casey, Mike Merchant, and Diane Silcox Reynolds. "Armyworm." AggieTurf.
aggieturf.tamu.edu/turfgrass-insects/armyworm

Sebudde, Rachel K., Richard Ancrum Walker, Tihomir Stucka, Pushina Kunda Ng'Andwe, Joseph Oryokot, Nkulumo Zinyengere, John Ilukor,

and Ashesh Prasann. *Uganda Economic Update 17th Edition: From Crisis to Green Resilient Growth – Investing in Sustainable Land Management and Climate Smart Agriculture (English).* Washington, D.C.: World Bank Group, 2021.
documents.worldbank.org/curated/en/265371623083730798/Uganda-Economic-Update-17th-Edition-From-Crisis-to-Green-Resilient-Growth-Investing-in-Sustainable-Land-Management-and-Climate-Smart-Agriculture

CHAPTER 3

Anyango, Chris. "Urbanization and Wellbeing in Africa." African Population and Health Research Center.
aphrc.org/runit/urbanization-and-wellbeing-in-africa

Button, Hannah. "Drought-Induced Loss of Livestock in Horn of Africa Will Impact Communities 'For Years to Come.'" Agrilinks. July 8, 2022.
agrilinks.org/post/drought-induced-loss-livestock-horn-africa-will-impact-communities-years-come

Dessie, Tadelle, and Okeyo Mwai, eds. *The Story of Cattle in Africa: Why Diversity Matters.* Nairobi, Kenya: International Livestock Research Institute, 2019.
cgspace.cgiar.org/handle/10568/108945

"Food System Vision Prize." Rockefeller Foundation.
rockefellerfoundation.org/initiative/food-system-vision-prize

Global Methane Hub.
globalmethanehub.org

International Livestock Research Institute.
ilri.org

Njarui, Donald M. G., Elias M. Gichangi, Mwangi Gatheru, Mupenzi Mutimura, and Sita R. Ghimire. Urochloa *(syn.* Brachiaria*) Grass Production Manual.* ILRI Manual 49. Nairobi, Kenya: International Livestock Research Institute, 2021.
cgspace.cgiar.org/bitstream/handle/10568/115848/Manual49.pdf?sequence=1&isAllowed=y

Njarui, Donald M. G., Elias M. Gichangi, Sita R. Ghimire, and Rahab W. Muinga, eds. *Climate Smart Brachiaria Grasses for Improving Livestock Production in East Africa – Kenya Experience: Proceedings of the Workshop Held in Naivasha, Kenya, 14–15 September, 2016.* Nairobi, Kenya: Kenya Agricultural and Livestock Research Organization, 2016.

kalro.org/sites/default/files/Proceeding-Climate-Smart-Brachiaria-Grasses-Dec2016.pdf

Reardon, Thomas. *Growing Food for Growing Cities.* Chicago Council on Global Affairs, 2016.
globalaffairs.org/research/report/growing-food-growing-cities

CHAPTER 4

Axmann, Heike. "SMP 2028 IndoDutch Taskforce Against Food Loss and Waste (INTAFLOW)." Wageningen University and Research.
research.wur.nl/en/projects/smp-2028-indodutch-taskforce-against-food-loss-and-waste-intaflow

Britannica. "Indo-Gangetic Plain." Last modified November 21, 2023.
britannica.com/place/Indo-Gangetic-Plain

"Changing Land Ownership, Agricultural, and Economic System." Environment and Society Portal.
environmentandsociety.org/exhibitions/famines-india/changing-land-ownership-agricultural-and-economic-systems

"Historical Changes in the Indo-Gangetic Plains of India." Gala.
learngala.com/cases/growing-into-an-uncertain-future/2

"How Is India Addressing Its Water Needs?" World Bank. February 14, 2023.
worldbank.org/en/country/india/brief/world-water-day-2022-how-india-is-addressing-its-water-needs

IPCC. "Sixth Assessment Report – Fact Sheet: Asia." October 2022.
ipcc.ch/report/ar6/wg2/downloads/outreach/IPCC_AR6_WGII_FactSheet_Asia.pdf

Mark, Joshua J. "Ancient India." World History Encyclopedia. November 13, 2012.
worldhistory.org/india

Mogno, Caterina, Paul Palmer, Christoph Knote, Fei Yao, and Timothy J. Wallington. "Seasonal Distribution and Drivers of Surface Fine Particulate Matter and Organic Aerosol Over the Indo-Gangetic Plain." *Atmos. Chem. Phys.* 21 (2021): 10881–10909.
doi.org/10.5194/acp-21-10881-2021

Naandi Foundation.
naandi.org

Sehgal Foundation.
smsfoundation.org

Shaw, Sameer H., Courtney Hammond Wagner, Udita Sanga, Hogeun Park, Lia Helena Monteiro de Lima Demange, Carolina Gueiros, and Meredith T. Niles. "Does Household Capital Mediate the Uptake of Agricultural Land, Crop, and Livestock Adaptations? Evidence From the Indo-Gangetic Plains (India)." *Front. Sustain. Food Syst.* (January 2019).
doi.org/10.3389/fsufs.2019.00001

"Taskforce Against Food Loss and Waste in India." Wageningen University and Research.
wur.nl/en/project/taskforce-against-food-loss-and-waste-in-india.htm

"The Story of India – Timeline." PBS.
pbs.org/thestoryofindia/timeline/1

U.S. Geological Survey. "New Map of Worldwide Croplands Supports Food and Water Security." November 14, 2017.
usgs.gov/news/featured-story/new-map-worldwide-croplands-supports-food-and-water-security

United Nations Environment Programme. "UNEP Food Waste Index Report 2021." March 4, 2021.
unep.org/resources/report/unep-food-waste-index-report-2021

Von Grebmer, Klaus, Jill Bernstein, Miriam Wiemers, Laura Reiner, Marilena Bachmeier, Asja Hanano, Olive Towey, Réiseal Ní Chéilleachair, Connell Foley, Seth Gitter, Grace Larocque, Heidi Fritschel, and Danielle Resnick. *2022 Global Hunger Index: Food Systems Transformation and Local Governance.* Bonn, Germany: Welthungerhilfe; and Dublin: Concern Worldwide; 2022.
globalhungerindex.org/pdf/en/2022.pdf

"Wheat Output in Africa and South Asia Will Suffer Severely from Climate Change by 2050, Modelling Study Shows." CIMMYT. June 16, 2023.
cimmyt.org/news/wheat-output-in-africa-and-south-asia-will-suffer-severely-from-climate-change-by-2050-modelling-study-shows

CHAPTER 5

Association of Guardians of the Native Potato of Peru. "History." Aguapan.
aguapan.org/en/historia

FAO Committee on Forestry. "State of the World's Forests 2016: Forests and Agriculture—Land Use Challenges and Opportunities." Food and Agriculture Organization of the United Nations. 2016.
fao.org/3/mq442e/mq442e.pdf

"Forest Pulse: The Latest on the World's Forests." World Resources Institute.
research.wri.org/gfr/latest-analysis-deforestation-trends

"RAP Llanos Se Consolida Como Despensa Agrícola de Colombia." Portafolio. August 18, 2021.
portafolio.co/economia/gobierno/rap-llanos-se-consolida-como-despensa-agricola-de-colombia-555263

Truitt Nakata, Ginya, and Margaret Zeigler. *The Next Global Breadbasket: How Latin America Can Feed the World: A Call to Action for Addressing Challenges & Developing Solutions.* Inter-American Development Bank, Global Harvest Initiative, 2014.
publications.iadb.org/en/next-global-breadbasket-how-latin-america-can-feed-world-call-action-addressing-challenges

USAID. "Growing Through Technology: Leveraging AgTech to Promote Prosperity in Guatemala." Medium. April 16, 2021.
medium.com/usaid-2030/growing-through-technology-751f7807a34a

Wight, Andrew J. "As Colombia Battles Deforestation, Other Post-Conflict Regions Show There's Hope for Conservation." NBC News. September 9, 2019.
nbcnews.com/news/latin-america/colombia-battles-deforestation-other-post-conflict-regions-show-there-s-n907811

CHAPTER 6

"About." Sustain-A-Grain.
sustainagrain.com/about

DeLonge, Marcia, and Karen Perry Stillerman. *Eroding the Future: How Soil Loss Threatens Farming and Our Food Supply.* Union of Concerned Scientists, 2020.
ucsusa.org/resources/eroding-future

King, Gilbert. "Where the Buffalo No Longer Roamed." *Smithsonian Magazine.* July 17, 2012.
smithsonianmag.com/history/where-the-buffalo-no-longer-roamed-3067904

Lakota Sioux Sičaŋǧu.
sicangu.co

Lev-Tov, Devorah. "Your Guide to Kernza: A Super Grain That's Good for You and the Planet." The Land Institute. January 25, 2022.

landinstitute.org/media-coverage/your-guide-to-kernza-a-super-grain-thats-good-for-you-and-the-planet

Najmabadi, Shannon. "Severe Drought Stunts Great Plains Wheat Crops Harvest in Nation's Breadbasket: Forecast to Be the Worst in 60-plus Years." *Wall Street Journal*. June 17, 2023.
wsj.com/articles/severe-drought-stunts-great-plains-wheat-crops-1b4d09a3

"Northern Great Plains." World Wildlife Fund.
worldwildlife.org/places/northern-great-plains

"People and Bison." National Park Service.
nps.gov/subjects/bison/people.htm

Scott, Michon. "National Climate Assessment: Great Plains' Ogallala Aquifer Drying Out." NOAA Climate.gov. February 19, 2019.
climate.gov/news-features/featured-images/national-climate-assessment-great-plains%E2%80%99-ogallala-aquifer-drying-out

"The Dust Bowl." National Drought Mitigation Center.
drought.unl.edu/dustbowl

Thompson, Matthew, dir. *Food 2050*. Episode 2, "7Gen." The Rockefeller Foundation, 2021.
food2050series.com/episodes/7gen

Wilson, Kelly, Stephanie Mercier, and Rob Myers. *Encouraging Farmer Adoption of Regenerative Agriculture Practices in the United States*. Chicago Council on Global Affairs, 2023.
globalaffairs.org/research/report/encouraging-farmer-adoption-regenerative-agriculture-practices-united-states

Wilson, Matthew. "Indigenous Paths Lead to a Green, Healthy Food System." Thomson Reuters Foundation. March 2, 2023.
context.news/nature/opinion/indigenous-paths-lead-to-a-green-healthy-food-system

World Wildlife Fund. *2022 Plowprint Report*. WWF, 2022.
worldwildlife.org/publications/2022-plowprint-report

ACKNOWLEDGMENTS

To ALL THE FARMERS and their families featured here who have boldly set out to grow against the grain of modern agriculture orthodoxy, I am deeply grateful. Thank you for sharing your remarkable endeavors and visions to hold off the Great Collision between humanity's two supreme imperatives: nourishing us all and preserving our planet.

This book has its roots in the work of the former Center on Global Food and Agriculture at the Chicago Council on Global Affairs. Many thanks to my colleagues for embracing and nurturing the project through the past few years: Peggy Tsai Yih, Alesha Miller, Ana Teasdale, Michael Kelley, Celine Ruby, Mignon Senuta, Brian Hanson, Marcus Glassman, Natalie Burdsall, Julia Whiting, Grace Burton, Gloria Dabek, Samanta Dunford, Mallory Hebert, Jody Oetzel, Vanessa Taylor, Laura Glenn O'Carroll, Kailey Griffith, and Catherine Bertini and Paul Schickler. And huge thanks as well for the astute research of Sarah Willis and Yasmine Adamali at Auburn University's College of Human Sciences and Hunger Solutions Institute.

As the scope of travel and reporting expanded, the Center